OPTIMIZATION IN PUBLIC TRANSPORTATION

Stop Location, Delay Management and Tariff Zone Design in a Public Transportation Network

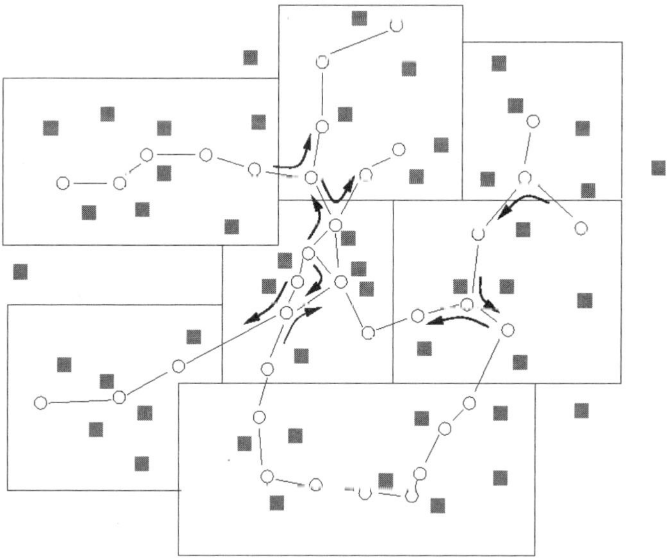

Optimization and Its Applications

VOLUME 3

Aims and Scope
Optimization has been expanding in all directions at an astonishing rate during the last few decades. New algorithmic and theoretical techniques have been developed, the diffusion into other disciplines has proceeded at a rapid pace, and our knowledge of all aspects of the field has grown even more profound. At the same time, one of the most striking trends in optimization is the constantly increasing emphasis on the interdisciplinary nature of the field. Optimization has been a basic tool in all areas of applied mathematics, engineering, medicine, economics and other sciences.

The series *Springer Optimization and Its Applications* publishes undergraduate and graduate textbooks, monographs and state-of-the-art expository works that focus on algorithms for solving optimization problems and also study applications involving such problems. Some of the topics covered include nonlinear optimization (convex and nonconvex), network flow problems, stochastic optimization, optimal control, discrete optimization, multi-objective programming, description of software packages, approximation techniques and heuristic approaches.

OPTIMIZATION IN PUBLIC TRANSPORTATION

Stop Location, Delay Management and Tariff Zone Design in a Public Transportation Network

By

ANITA SCHÖBEL
Georg-August University, Göttingen, Germany

Springer

Library of Congress Control Number: 2006929191

ISBN-10: 0-387-32896-3 e-ISBN: 0-387-36643-1

ISBN-13: 978-0-387-32896-6

Printed on acid-free paper.

AMS Subject Classifications: 90B06, 90C10, 90C90, 90C27, 90B85

Printed in the United States of America.

9 8 7 6 5 4 3 2 1

springer.com

To my parents
Helga and Volker Schumacher

Preface

Der Kunde ist König.

(German saying)

Public transportation plays an important role in most populated areas. Especially in metropolitan regions public transportation systems are widely used. But unfortunately, public transportation is often a subject of complaints. Customers are annoyed about "unfair prices", about "bad service" and in particular get upset in case of delays. Such complaints are understandable, but for the public transportation companies it is often impossible to provide a better service without increasing the costs. The reason for these difficulties is the complexity and the size of the planning problems arising.
The theory of optimization provides a sound methodology for finding good solutions, if a mathematical model of the respective problem is known. Moreover, due to the availability of fast computers many problems that seemed to be intractable some years ago can nowadays be solved.

This work provides suitable models for planning public transportation systems from a customer-oriented point of view, but taking into account the limited budget public transportation companies have to respect. In particular, we develop and analyze optimization models for the following three problems:

Part I: Stop location. Here we deal with the location of stops along bus routes, or of stations along railway tracks. As objective functions we consider the number of customers living close to a station and the additional travel time arising by the stopping activities of the trains or buses. In particular, we discuss how to find the minimal number of stops to cover a given set of demand points or demand regions, how to cover as many customers as possible with a given budget and both problems together in a bicriteria setting.

Part II: Delay management. If a vehicle arrives at a station with a delay, passengers who wish to change into another vehicle, say a bus, may miss their connection, if this bus departs on time. Such wait-depart decisions and their impact on the whole transportation system are investigated from the customers' perspective. As objective functions we hence discuss the sum of all delays over all passengers, the number of missed connections, and the sum of all delays over all vehicles. The latter two objectives are treated as a bicriteria optimization problem.

Part III: Zone planning. In order to design a zone tariff system, the complete transportation area has to be partitioned into zones, and prices for traveling through 1,2,3,... zones have to be defined in such a way that the current income of the public transportation company does not decrease too much. As objective function we consider the deviations between the new prices and some given reference prices. These deviations can be interpreted as the fairness of the new tariff system or as the changes to the current ticket prices.

All three problems were brought to my attention within real-world projects, and some of the obtained results have already been implemented and applied in practice. Nevertheless, the main focus of this work is to develop a consistent mathematical theory and to present basic results within all three fields.

- The stop location problem is treated using the concept of gauges and ideas of continuous location theory. A finite dominating set of possible new stops can be derived. This allows us to formulate the stop location problem as a set covering problem. By using the special structure of the covering matrix which is due to the geometrical properties of the stop location problem, efficient solution methods for this type of set covering problems are developed.
- For the delay management problem three different, but equivalent mixed integer programming formulations are presented. By combining these models many structural results for the delay management problem are obtained. In particular, it is possible to identify cases in which the problem is solvable efficiently. Furthermore, methods of project planning are applied to determine Pareto solutions.
- Finally, the design of zone tariff systems in public transportation is modeled by methods of graph theory. The obtained theoretical results together with ideas of clustering theory are utilized for deriving solution approaches.

The theory presented in this text and the obtained results open a wide field for further developments and implementations of the suggested approaches. The algorithms that have already been tested on our real-world data confirm the practical usefulness of the models and show their potential for future applications.

Before concluding the preface I wish to add several acknowledgments. First of all, I thank Horst W. Hamacher for his support, for the pleasant and constructive work together with him, and for his helpful advice in any question I had.

For many valuable suggestions I am indebted to Kathrin Klamroth and Dagmar Tenfelde-Podehl. I also appreciate the comments of Teresa Melo, Martin C. Müller, and Michael Schröder.

It was a pleasure to work with Annegret Liebers and Dorothea Wagner on the stop location problem. My thanks also go to Frank Geraets of Deutsche Bahn who provided this nice problem together with real-world data. I want to thank Andreas Ginkel and Nikolaus Ruf for many fruitful discussions we had during the preparation of their diploma theses on delay management and set covering problems. Moreover, I want to express my gratitude to Dieter Grünewald whose expertise in analyzing zone plans influenced my practical work in this area.

I also want to thank Robert Saley of Springer for his kind assistance during the publication of the manuscript.

Last, but not least, my special thanks go to my husband Georg and to my children Svenja and Malte for their encouragement, patience, and understanding which made it possible for me to write this text.

Göttingen,
December 2005

Anita Schöbel

Contents

Part II Delay Management

Part III Tariff Planning

1

Customer-oriented Traffic Planning

1.1 Customer-oriented Transportation

Although public passenger transportation plays an important role especially in large metropolitan areas, it also has to be carefully planned and organized in a rural environment. There are economical, environmental, and social reasons for considering the needs of customers when planning public transportation.

- First of all, if a public transportation company attracts more customers then it will sell more tickets and hence its income will (usually) increase.
- An environmental aim is to decrease the amount of individual traffic (mainly in large cities) and thus reduce its negative effects such as pollution, noise, and congestion. This is sometimes accomplished by imposing restrictions or fines through high parking fees, tolls, closure of roads, or car-free days. A way of avoiding this would be to offer such a good alternative that (at least some) people voluntarily decide to use public transportation instead of their cars.
- In areas with few inhabitants, congestion usually is no problem. The challenge here is to offer an affordable transportation mode for people who do not have the opportunity to travel by car, e.g., children, elderly people, or citizens without a driving license.

We now briefly introduce in an informal way some of the problems occurring in public transportation. Three of them – locating stops, delay management, and tariff planning – will be discussed in detail in subsequent chapters. An overview of the problems considered in this text and their relation to other customer-related steps in the planning phase and at the operational level is given in Figure 1.1. Note that sometimes the same model can be used for on-line decisions and for long-term decisions at the same time. This is for example the case for the model that we will present for the delay management problem.

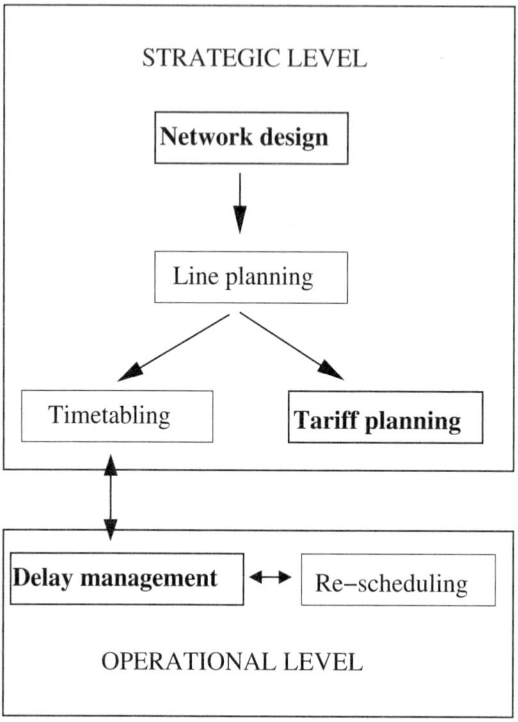

Fig. 1.1. The problems considered in the text within the planning process in public transportation.

Network planning

Network planning includes the design of the transportation network, i.e., siting the stations and the bus routes or train tracks. The outcome of the process of network planning is the *public transportation network* (PTN). Network planning problems have been treated in the general context of network flow problems. In the public transportation literature we refer, e.g., to [CW86, BM95, CG02] and to the references given therein. However, in real life a PTN is usually not designed from scratch, but only modifications of an existing PTN are considered, such as

- finding new stations in a railway or bus network,
- closing existing stations, or
- finding a subnetwork for opening rapid transit lines.

For these problems literature is rather sparse. Locating stops or stations in the PTN will be discussed in Part I of this text. For finding subnetworks for operating an underground system or a rapid transit line, hub location models have been developed by [NSS01].

Line planning

Line planning concerns the definition of paths in the PTN on which service should be offered, i.e., the routes of the bus or railway lines. The line planning problem has been well studied in the literature. For an early contribution we refer to [Die78]. In [BKZ96, Bus97] the goal is to maximize the number of passengers with direct connections under the constraint that all passengers can be transported. The solution methods proposed use advanced integer programming techniques. Under a similar constraint, the goal in [CvDZ96] is minimization of costs for the public transportation company. Line planning problems considering different types of vehicles simultaneously were studied in [GvHK04, GvHK02]. Various models and algorithms are discussed in [Goo04].

A new approach is to take into account that the behavior of the customers depends on the design of the lines. A first model including such demand changes was treated with simulated annealing in cooperation with *Deutsche Bahn*, see [Kli00b, Sch01a]. Moreover, the choice of the routes of customers depends on the (unknown) line plan. Finding a line plan together with optimal routes for the customers has recently been considered in [SS05, Sch05b, BGP04a, BGP04b, Sch05a, LMMO06]. In these approaches, the goal is to design lines in such a way that the traveling time of the customers is minimized. The first two of these publications also include the number of transfers of customers in the objective function. The special case of locating one single line so as to maximize the number of passengers is treated in [LMO05].

Timetabling

Timetabling determines the departure and arrival times for all trips at all stations. Here two cases are distinguished.

Case 1: All rides within the same line start in *periodic* time intervals, e.g., at 7:03, 7:33, 8:03, 8:33, 9:03 and so on.
Case 2: The timetable of the rides is *non-periodic*.

Many papers and theses deal with problems related to timetabling. An overview of the literature in this area will be given in Section 6.2 (see page 98) within the context of delay management.

Tariff planning

Tariff planning concerns the determination of fares for the customers. Different systems are possible.

distance tariff: The price depends on the length of the journey.
unit tariff: Each journey costs the same.

zone tariff: The complete area is partitioned into zones, and the prices depend on the number of passed zones, from the origin to the destination of the journey.

In tariff planning the problem is to design a new tariff system along with its prices. A common requirement is that the new income of the public transportation company should not decrease compared to its current income. On the other hand, the customers should find the new system acceptable. Designing zone tariffs under such criteria will be treated in Part III.

Summarizing, line planning, timetabling and network design problems have been well studied in the literature so far. There are other problems belonging to the strategic planning process in public passenger transportation, like

- rolling stock circulation,
- vehicle scheduling,
- shunting,
- crew management,
- crew rostering,
- maintenance issues.

Since these problems have no direct effects on the customers they will not be considered in this text. Various models and solution approaches for these problems exist. For references the reader is referred, e.g., to the proceedings of the CASPT meetings which are mentioned below.

We finally list two operational problems, which have to be solved on-line in case of disturbances.

Delay management:

Suppose that a vehicle arrives at a station with a delay. Should a connecting vehicle wait for passengers who wish to change or should it depart on time? The goal is to minimize the inconvenience caused by delays from the customers' point of view. The delay management problem will be treated in Part II.

Re-scheduling of vehicles:

Especially in rail transportation, construction sites, delays, or any other disturbance make a re-scheduling of trains necessary. This is a difficult problem since many constraints have to be taken into account. The main requirement in many railway companies is that no two trains are allowed to occupy the same segment of a track (called *block*) at the same time. The goal may be to return to the original schedule as quickly as possible, or to minimize the additional delays of the trains. The problem has mainly been considered in transportation and engineering sciences, and is practically often solved based on priority rules. Operations research models can be found in [AFT02, BHK99, ADGGT99, Kro97, Zwa96, AD96], while there are also many other successful approaches from various areas, including [Tör05b, TJ05, WS05, PMP04, Jac04, vE01, Fay00, HKF96, PT82]. A recent overview with many references is given in [Tör05a].

Other operational problems include the re-scheduling of crew in case of unexpected absent drivers, re-planning of rosters, or maintenance re-scheduling.

For more details about the mathematical models used in the planning process in public transportation we refer to the basic rail transportation models of [Ass80] and to the survey of Bussieck, Winter and Zimmermann [BWZ97]. Another overview is given by Borndörfer, Grötschel and Löbel [BGL98]. Patriksson and Labbé [PL02] collected articles about the state-of-the-art in the field of transportation planning. The survey of Cordeau, Toth and Vigo [CTV98] focuses in particular on routing and scheduling in rail transportation. Railway planning problems are also addressed in the surveys of [Wag03, GJP+04] and in the forthcoming collection [GKS+06].
Moreover, we refer to the conference proceedings of the CASPT (Computer-Aided Scheduling of Public Transport) meetings [Wre81, Rou85, DW88, DR92, DBP95, Wil99, VD01], to the TRISTAN (Triennial Symposium on Transportation Analysis) [BT96] conferences, and to the proceedings of the ATMOS workshops [Zar01, Wag02, Ger04, GKS+06].

1.2 Public Transportation Network and Customer Data

We start with a formal definition of a public transportation network, a simple example of which is depicted in Figure 1.2.

Definition 1.1. *A* **public transportation network** *is a finite, undirected graph* PTN $= (V, E)$ *with*

- *a node set V representing stops or stations, and*
- *an edge set E, where each edge $e = \{u, v\}$ indicates that there exists a direct ride from station u to station v (i.e., a ride that does not pass any other station in between).*

In public transportation, an ordered pair of stops (or stations) u, v is often called a *relation*.

Within the network design step, the PTN is constructed, or modifications of an already existing PTN such as adding new stops or closing existing ones are planned. However, for all other purposes, like line planning, timetabling, delay management, or tariff planning, we assume the PTN as given and fixed. It may happen that the set E of direct rides in the PTN is not given, but a

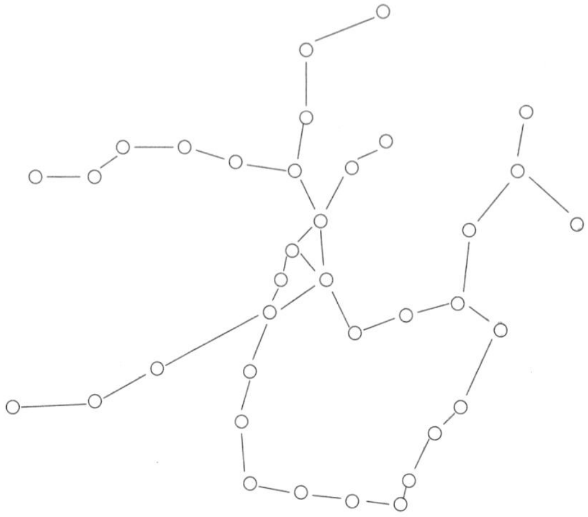

Fig. 1.2. A PTN with its set of stops V and its direct rides E.

timetable is at hand. Possibilities to construct E in such a case are discussed in [Lie01].

Since we mainly deal with optimization problems from the customers' point of view, we now discuss the data about the customers needed for our models.

OD-matrix: The origin-destination matrix (OD-matrix) $W = (W_{uv})$ is a $|V| \times |V|$ matrix containing the number of customers who wish to travel from station u to station v for all relations (u, v) in the PTN. Instead of the number of customers, the number of sold tickets can be given. The latter is in particular needed for tariff planning.

Traffic load: The traffic load is defined by the number of customers traveling along an edge $e \in E$ or through a node $v \in V$ in the PTN, and is denoted by c_e or c_v, respectively. The traffic load can be given as number of customers per hour, per day, per week, or per year. The traffic load will be used for the stop location problem (Part I), and as an approximation in the delay management problem (in Part II).

We now summarize some notation that will be used throughout the text.

Notation 1.2. *Let* PTN $= (V, E)$ *and let* I *be a fixed time interval.*

- *For all* $u, v \in V$ *let* W_{uv} *denote the* **demand** *of relation* (u, v), *i.e., the number of passengers who wish to travel from station* u *to station* v *within the time interval* I. *The matrix* $W = (w_{uv})_{u,v \in V}$ *is called the* **origin-destination matrix** *or, shorter, the* **OD-matrix**.
- *For all* $e \in E$ *let* c_e *denote the* **traffic load of an edge** e, *i.e., the number of customers using edge* e *within the time interval* I.

Moreover, for $v \in V$ let c_v denote the **traffic load of station** v*, i.e., the number of customers traveling through station v within the time interval I.*

Note that I needs to be chosen appropriately for the respective application one has in mind. For example, in tariff planning, I usually refers to a long period such as a whole year, hence W_{uv} can be used to calculate the annual income on the relation from u to v. For line planning, however, the traffic load is important to make sure that all passengers can be transported. Here I is usually a short interval, like the morning traffic period (e.g., from 6 to 8 a.m.), and c_e is used to calculate the number of vehicles needed along edge e within this period. Apart from these widely used data we sometimes need the following more detailed information about the customers.

Demand within a point or region (needed in Part I): When dealing with the location of stops close to customers, we assume that a set of demand points or demand regions is given. The number of (potential) customers within a demand point d is denoted by w_d. Alternatively, the number of (potential) customers within a demand region D is called w_D.

Paths of the customers (needed in Part II): For calculating the delay of a customer, it is not enough to know where his journey has started and to which station he wishes to travel. Also of interest are the starting time and the stations in which a transfer to other vehicles occurs, i.e., the path followed within the transportation network, as well as the vehicles used. For a path p let w_p denote the number of customers using this path. Since the information about such paths is often not available we will also present models which do not rely on this specific information.

Number of changing passengers (needed in Part II): For calculating how many passengers miss a connection a from some vehicle g to another vehicle h at a station v we need the number of transfer passengers w_a who plan to use connection a to change between the respective trains.

Destinations of the customers (needed in Part II): In delay management, it is also convenient to use the number of customers C_v^g who reach their final destination v traveling in some vehicle g.

Note that the latter two data sets can be easily obtained if the paths of the customers are known. If only the OD-matrix W is known, it is possible to approximate the traffic loads by finding a set of reasonable paths from u to v for each relation u, v and dividing W_{uv} among these paths. Doing this for all relations and then adding for each edge $e \in E$ the weights of all paths containing e gives an approximation of the traffic load of e. Formally, this is stated next.

Algorithm 1: Approximating traffic loads

```
Input: PTN and OD-matrix W = (W_uv)_{u,v∈V}.
Output: Traffic load c_e for each edge e ∈ E.
Step 1. For each pair u,v ∈ V with W_uv > 0 determine a set of ``reasonable''
   paths from u to v
```

$$P_1^{uv}, P_2^{uv}, \ldots, P_k^{uv},$$

and assign weights $w_{P_1^{uv}}, w_{P_2^{uv}}, \ldots w_{P_k^{uv}}$ to these paths in such a way that

$$\sum_{i=1}^{k} w_{P_i^{uv}} = W_{uv}.$$

Step 2. For all $e \in E$ set

$$c_e = \sum_{u,v \in V} \sum_{\substack{i=1,\ldots,k: \\ e \in P_i^{uv}}} w_{P_i^{uv}}.$$

Note that the difficulty of the algorithm above is to express the customers' behavior by an appropriate set of weighted paths, i.e., the skills are more of a practical nature rather than of mathematical hardness. For simplicity, k is set to 1 in many applications, and the only path P^{uv} for the relation from u to v is chosen as a shortest path. In this case, Algorithm 1 simplifies to computing the following expression for all $e \in E$:

$$c_e = \sum_{u,v \in V : e \in P_{uv}} W_{uv}.$$

Part I

Stop Location

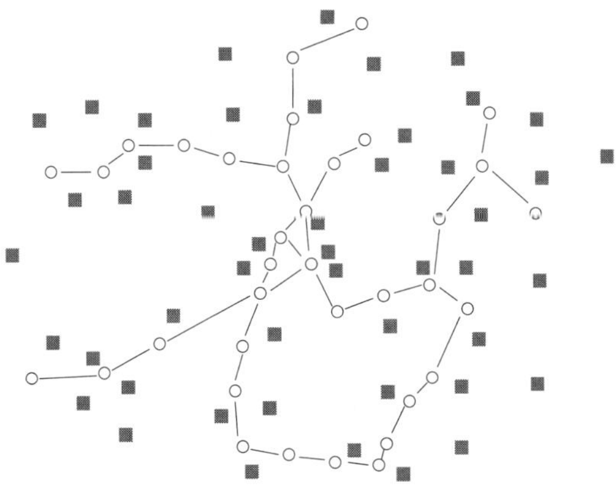

2

Introduction

Establishing stops (or stations) within a transportation network is fundamental for offering public transportation service, since stops are an important part of the PTN. But it is not clear in advance, how many stops are reasonable, and where they should be built. Let us consider the effects of stops on the customers:

- On the one hand, many stops are advantageous from the customers' point of view, since they increase the accessibility of the trains or buses. Establishing a new stop may hence attract new customers and increase the demand. In bus transportation, the *covering radius* is often assumed to be 400 m, meaning that a customer will think about using a bus, only if the next bus stop is within a distance of at most 400 m. In rail transportation, the covering radius is larger, and is usually assumed to be 2 km.
- On the other hand, each additional stop increases the transportation time (e.g., by two minutes in rail transportation) for all trains or buses stopping there. This makes the transportation service unattractive to customers.

Moreover, this additional running time of trains (or buses) is costly for the transportation company, and also fixed costs arise for establishing a new stop.

In the *continuous stop location problem* we deal with the location of new stops along a given track system. This means, we assume that the tracks for the trains are already built, or the routes for the buses are already fixed. For the sake of simplicity we will use the wordings "stops" and "tracks" in the following, but keep in mind that the models and algorithms presented can also be applied for bus transportation.

We further assume a (possibly empty) set of already existing stops or stations. As input data we also need the locations of the potential customers, given as points or as regions in the plane, and the traffic load along the edges of the given tracks. An example for a set of demand points is depicted in Figure 2.1. Our goal is to locate additional stops along the tracks such that

- as many (potential) customers as possible live closer than a given radius r to their nearest stop, and such that
- the increase of travel time caused by the new stops is as small as possible.

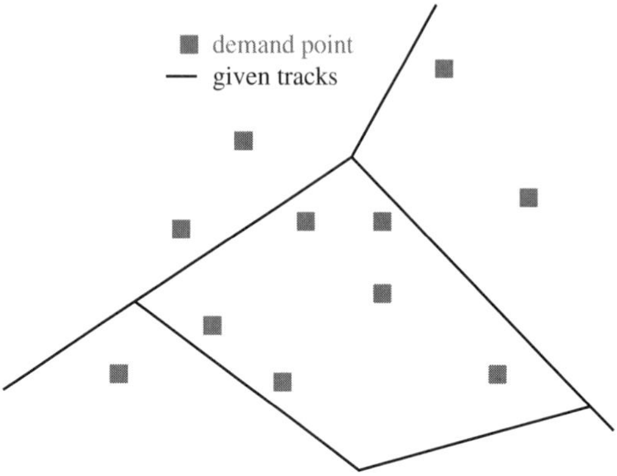

Fig. 2.1. The set of tracks $\mathcal{T}$ and a set of demand points $\mathcal{D}$ in the plane.

The result we obtain by solving the continuous stop location problem defines the PTN which is the basis for many subsequent optimization models in public transportation planning. Establishing no stop at all means that the additional travel time is minimal, but for none of the customers does the accessibility increase. The other extreme is to open stops until the complete demand is covered. The following optimization problems will be treated in this chapter.

- In the *complete cover stop location problem (CSL)* we want to cover all potential customers with as few stops as possible, or with as few costs as possible. The problem will be treated in Chapter 3 for the case that the demand is given at points and in Section 5.1 for the case of demand regions.
- The *bicriteria stop location problem (BSL)* focuses on minimizing the additional travel time and on maximizing the covered demand simultaneously. This provides solutions between the two extremes of covering the complete demand and of establishing no (additional) stop at all. (BSL) is discussed in Chapter 4.
- In the *door-to-door travel time stop location problem (DSL)* we investigate the door-to-door travel time over all customers. The door-to-door travel time for a customer is given by the time he needs to get to the first station of his trip plus the time of the trip itself plus the time he needs to reach his final destination after leaving the public transportation system. (DSL) will be considered in Section 5.2.

Chapter 2 is structured as follows: We start by presenting the application which motivated us to deal with continuous stop location problems. A literature review on stop location is given next. Then we present a model for the continuous stop location problem, enabling us to evaluate the interesting objective functions.

2.1 Application

When comparing railway systems all over Europe, it turns out that Switzerland has a higher amount of rail transportation than other countries. Among others, one reason could be that in Switzerland the number of stops compared to the overall length of the track system is significantly higher than in other countries. The interesting question arising by this observation is, if it is an advantage or a disadvantage to have many stops. To come to an answer, we consider a customer-oriented point of view. A quality criterion for the customers which is influenced by the number of stops is the door-to-door travel time of their journeys, including the time they need to get from home to their departure stations and the time they need to reach their final destinations. A priori it is not clear if this time will increase or decrease by opening new stops along the track system.

Note that by a stop we do not mean a fully equipped station, but just a stopping point for the trains, which is relatively cheap for the railway company. Our results and some of our algorithmic approaches have been implemented and tested using data of the largest German railway company, *Deutsche Bahn*. Here we located new stops along the track system, relevant for regional trains, i.e., all regional trains are supposed to stop while the fast long-distance trains pass through. Our real-world data is described next.

- We use 30 637 demand regions, given as polygons with an average of 45 nodes per polygon. These polygons are not identical with the borders of the communities and also do not form a partition of Germany. They represent the population distribution better than community borders since green land is excluded. This means that most of the data is very accurate; even relatively small towns are given as a set of more than 10 different demand regions.
- The PTN we used represents the network of *Deutsche Bahn*. It has a size of 6 828 stations and 8 724 edges.
- For each demand region we furthermore know the number of inhabitants, and for each edge we got an approximation of the traffic load, i.e., the number of customers using the edge.

Moreover, *Deutsche Bahn* specified some of the necessary parameters for our models. The time needed for an additional stopping activity of a regional train was estimated as two minutes. For the covering radius, a distance of 2 km is often used in rail transportation.

2.2 Literature Review

The importance of planning stops carefully and different customer-oriented criteria for bus stop location were already discussed in the case study of Demetsky et al, see [DAL82]. Among the many possible objective functions one goal is to establish as few stops as possible in such a way that all customers are covered. This was done in [Gle75] and in [MDSF98, Mur01a, Mur01b]. In the latter papers, the public transportation network in Brisbane, Australia was analyzed in detail and it turned out that 84.5 % of the stops are not necessary in terms of covering a set of given demand points within a Euclidean distance of 400 m, i.e., closing them would not decrease the actual number of covered customers. The stop location problem was treated in a discrete setting in these papers, i.e., the authors either considered only the actual stops, or they assumed that a finite candidate set of new stops is given. This leads to an unweighted set covering problem, also called *location set covering problem* which was introduced in [TSRB71, TR73]. In the context of stop location this problem has been solved by [Mur01a] using the Lagrangian-based set covering heuristic of [CFT99]. A new discrete stop location model was developed by Laporte et al. [LMO02]. They investigate which candidate stops along one given line in Seville should be opened, taking into account demand regions and constraints on the inter-station space. The coverage of a new stop is determined using a gravitation model. Finally, they solved the problem by a longest path algorithm in an acyclic graph. Their model resembles the *maximum coverage location problem* originally presented in [CR74, WC74].

The difference between the continuous stop location problem considered here and most papers published so far is that in the continuous stop location problem we do *not* choose the stops from a known set of possible candidates, but allow establishing a new stop anywhere along the given railway tracks (or along the given bus routes). The covering information can hence not be given explicitly but must be calculated by some (geometric) formula. The first approaches dealing with a continuous candidate set were given in [HLS$^+$01] and [SHLW02]. They are described in more detail in Section 5.2 and in Chapter 3. The results of [RS04, Sch05c, SS03] are based on these two papers and can be found in Section 3.6, Chapter 9, and Section 5.1. The research of [KPS$^+$03] was also motivated by this research. They deal with a variant of the continuous stop location problem, aiming to cover as much demand as possible with a given number of new stops, see Section 4.1. In [MMW04] the stop location problem has been investigated and solved for the case of two intersecting lines. Solving the stop location problem by data reduction of the underlying covering problem has been studied in [Mec03] and in [MW04].

2.3 A Model for Continuous Stop Location

Let $G = (V, E)$ be a finite, planar graph with straight-line embedding in the plane. In real-world data sets, the nodes of G represent either existing stations or important breakpoints. We identify each edge $e \in E$ by a line segment in the plane. Moreover,

- c_e is the traffic load along edge $e \in E$, i.e., the number of customers using edge e, and
- c_v is the traffic load through station $v \in V$, i.e., the number of customers passing through station v (and not getting on or off there).

Both parameters can be given, for example, in customers per day.

Definition 2.1. *Given $G = (V, E)$ define the* **track system**

$$\mathcal{T} = \bigcup_{e \in E} e = \{x \in \mathbb{R}^2 : x \in e \text{ for some } e \in E\} \subseteq \mathbb{R}^2$$

as the set of points on edges of the planar embedding of G.

Our goal is to establish stops (or stations), which are represented by points in $\mathcal{T}$. The evaluation of a set $S \subseteq \mathcal{T}$ is described next.

Additional Travel Time

To calculate the additional travel time induced by some set of stations $S \subseteq \mathcal{T}$ we take the number of customers affected by the additional stopping activities and multiply them by the time t_{stop} which is needed for an additional stop. According to *Deutsche Bahn*, t_{stop} can be assumed to be two minutes, independent of the location of the stop. This is specified in the following notation:

Definition 2.2. *Given $s \in \mathcal{T}$ let*

$$g(s) = \begin{cases} s \text{ if } s \in V \\ e \text{ if } s \in e, s \notin V. \end{cases}$$

Furthermore, given a finite set $S \subseteq \mathcal{T}$ we define

$$f_{\text{time}}(S) = \sum_{s \in S} t_{\text{stop}} c_{g(s)}.$$

For an infinite set S we define $f_{\text{time}}(S) = \infty$.

Since t_{stop} is a constant, e.g., two minutes in rail transportation, it can be neglected for the optimization process. Furthermore, note that $f_{\text{time}}(S) = |S|$ if all traffic loads $c_{g(s)} = 1$, i.e., if we assume that each edge is used by exactly one customer. Hence, we will refer to the *unweighted problem* if we deal with the special case of minimizing the number of stations.

The Cover of a Set of Stops

To deal with the accessibility of potential customers, we next assume that $\mathcal{D} \subseteq \mathbb{R}^2$ is a finite set of either

- demand points, or of
- pairwise disjoint demand regions

representing important points or regions such as settlements, industrial areas, shopping centers, or leisure parks.

Notation 2.3. *For $\mathcal{D}$ let*

$$D_{total} = \bigcup_{D \in \mathcal{D}} D$$

be the **demand set**. *Note that $D_{total} = \mathcal{D}$ if $\mathcal{D}$ consists of demand points.*

We now introduce the notion of *covering* with respect to a distance measure γ. We may specify different distance measures for each of the elements of $\mathcal{D}$, i.e., for each of the demand points or regions. As distance measure γ_D we allow any norm or gauge (see Appendix C); readers who are not familiar with gauges may simply imagine γ_D as the Euclidean distance. For $d \in \mathbb{R}^2$, $S \subseteq \mathbb{R}^2$, let (as usual)

$$\gamma_d(d, S) = \min_{s \in S} \gamma_d(d, s).$$

Notation 2.4. *Let $d \in D_{total}$. Then γ_d denotes the distance measure associated with d.*

If $\mathcal{D}$ consists of demand regions, and $D \in \mathcal{D}$, then we require for all points $d_1, d_2 \in D$:

$$\gamma_{d_1} = \gamma_{d_2} = \gamma_D.$$

A demand point is covered, if the distance to its closest station is smaller than or equal to a given radius r, where the used distance need not be the same for all demand points. Formally, this is specified below.

Definition 2.5. *Given $r > 0$, and $S \subseteq \mathcal{T}$.*

1. *A point $d \in D_{total}$ is* **covered** *by S if $\gamma_d(d, S) \leq r$.*
2. *Furthermore, the* **cover** *is S is $\text{cover}_{\mathcal{D}}(S) = \{d \in D_{total} : d \text{ is covered by } S\}$.*

If it is clear to which set $\mathcal{D}$ we refer, we just write $\text{cover}(S)$. Furthermore, for $s \in S$ we use $\text{cover}(s)$ for $\text{cover}(\{s\})$. Note that if $\gamma_d = \gamma$ for all $d \in D_{total}$ we obtain

$$\text{cover}_{\mathcal{D}}(S) = \{d \in \mathbb{R}^2 : \gamma(d, S) \leq r\} \cap D_{total}.$$

The cover of a point is illustrated in Figure 2.2. The small rectangles in parts (a) and (b) represent the demand points $d_1, \ldots, d_6$, while we consider two demand regions D_1 and D_2 in parts (c) and (d). All elements of $\mathcal{D}$ in parts (a) and (c) are assumed to have the Euclidean distance associated with them. In part (b), $\gamma_{d_1}, \gamma_{d_2}$, and γ_{d_3} equal the rectangular distance, while the remaining elements $d \in D_{total}$ again have γ_d as Euclidean distance. In part (d), we assume γ_{D_1} as rectangular distance and γ_{D_2} as Euclidean. In parts (a) and (b) the cover consists of the filled small rectangles, in parts (c) and (d) the cover is given by the dashed area.

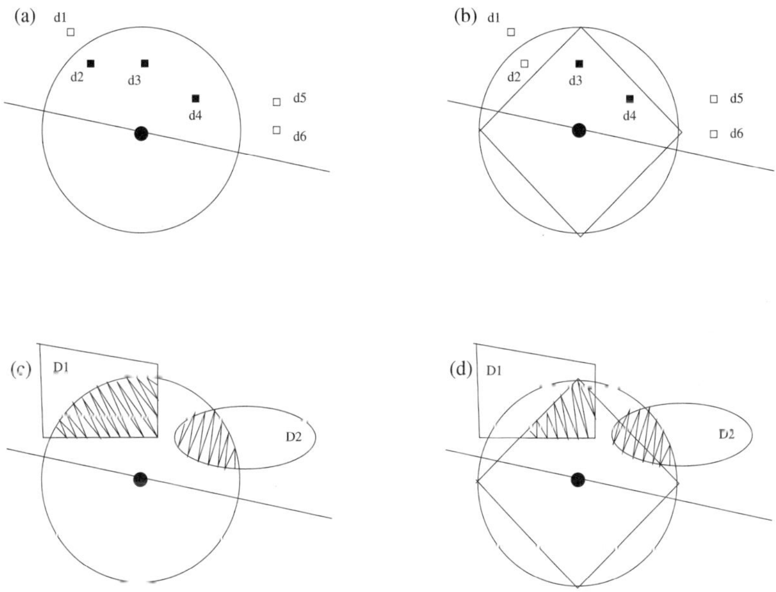

Fig. 2.2. The cover for demand points (see (a) and (b)) and for demand regions (in (c) and (d)), both for the Euclidean distance (see (a) and (c)) and for mixed rectangular and Euclidean distances (in (b) and (d)).

We further need the following notation. Consider $d \in D_{total}$ with associated distance function γ_d. Let $B_d = \{x \in \mathbb{R}^2 : \gamma_d(x) \leq 1\}$ be the unit ball associated with γ_d, see Appendix C. Using the denotation

$$B_d^r = d + rB_d,$$

we get

$$\gamma_d(d, x) \leq r \text{ if and only if } x \in B_d^r.$$

Hence, we obtain:

Lemma 2.6. *Let $d \in D_{total}$ and $S \subseteq \mathcal{T}$. Then d is covered by S if and only if $S \cap B_d^r \neq \emptyset$.*

We refer to Figure 2.3 for an illustration.

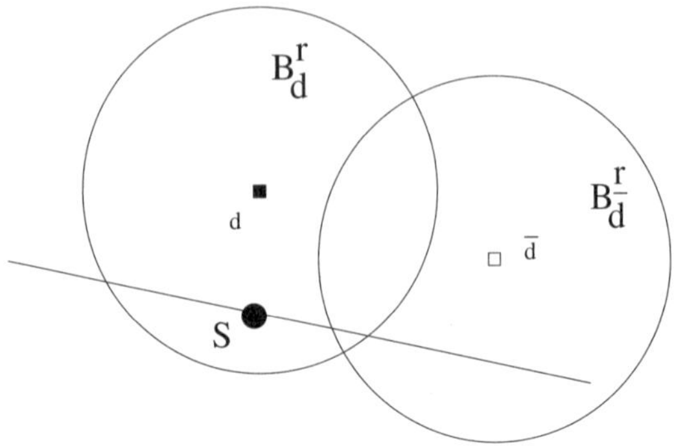

Fig. 2.3. $B_d^r \cap S \neq \emptyset$, hence d is covered by S. On the other hand, $B_{\bar{d}}^r \cap S = \emptyset$, hence $\bar{d}$ is not covered by S.

We will often use this *dual view* of the stop location problem, not considering the cover of some points $S \subseteq \mathcal{T}$ but starting from one point $d \in D_{total}$. For $d \in D_{total}$ we determine the set of points on $\mathcal{T}$ which can be used to cover d, i.e., those points where the location of a new stop would attract the demand in d.

Notation 2.7. *Let* $d \in D_{total}$. *Then* $\mathcal{T}(d) = \{s \in \mathcal{T} : \gamma_d(d,s) \leq r\}$.

$\mathcal{T}(d)$ can be calculated by intersecting the unit ball B_d^r of the gauge γ_d (with radius r) centered at the demand point d with the set of tracks $\mathcal{T}$, as the following lemma shows. For an illustration, see Figure 2.4.

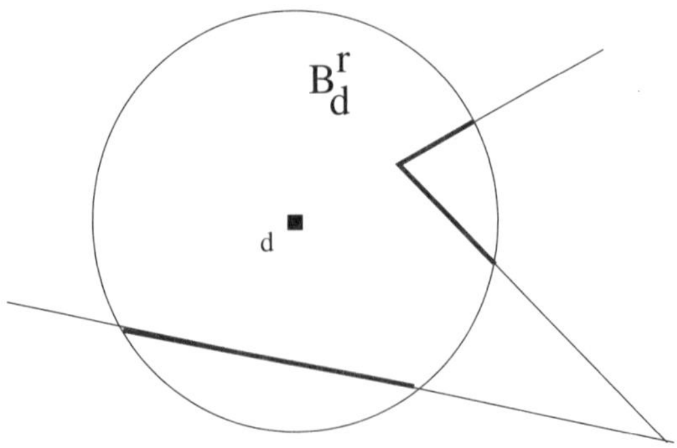

Fig. 2.4. The set $\mathcal{T}(d) = B_d^r \cap \mathcal{T}$ (the thick part of the tracks).

3

Covering All Demand Points

Throughout this chapter let $\mathcal{D} \subseteq \mathrm{IR}^2$ be a finite set of points in the plane, i.e., each $d \in \mathcal{D}$ is a point given by its geographic coordinates $d = (d_1, d_2)$. These points may represent larger demand regions, a simplification which is used in almost all papers about stop location (except [LMO02]). Note that in Section 5.1 we will extend the methods of this chapter to the case of demand regions.
Our goal in this chapter is to cover all of the given demand points. We refer to this problem as *complete stop location problem* (CSL). Note that Lemma 2.10 shows that we need not deal with already existing stations as in (SL'), but can plan all stations from scratch in (CSL). The problem we consider here hence is the following.

(CSL)

Given $G = (V, E)$ with its set of points $\mathcal{T} = \bigcup_{e \in E} c$, traffic loads c_e for all $e \in E$, and c_v for all $v \in V$, and a finite set of points $\mathcal{D}$ with gauges γ_d for all $d \in \mathcal{D}$, find a set $S \subseteq \mathcal{T}$ covering all points in $\mathcal{D}$ such that the additional travel time

$$f_{\text{time}}(S) = \sum_{s \in S} c_{g(s)}$$

is minimized.

Recall that $g(s)$ is the edge or the node, respectively, where stop s is located, and $c_{g(s)}$ is the traffic load, i.e., the number of customers passing through the edge, or the node, respectively.

Chapter 3 is structured as follows: In this chapter, we first discuss some general properties of the complete continuous stop location problem, such as feasibility and the NP-hardness of the problem. We then present a finite set of points $\mathcal{S}$ on the tracks for which we can show that they always contain an optimal solution. Using this finite dominating set, the continuous stop location problem can be transformed to a discrete set covering problem. Instead of using set covering approaches from the literature, we investigate the structure of the covering matrix. We identify cases in which this matrix has the consecutive ones property. Furthermore, we present decomposition approaches. Finally, we develop efficient solution approaches for the stop location problem in a more general context, namely for *set covering problems* in which the coefficient matrix (almost) has the consecutive ones property.

3.1 Feasibility and Complexity of Complete Cover

Problem (CSL) can be summarized as

$$\min f_{\text{time}}(S)$$

$$\begin{aligned} \text{such that} \quad \text{cover}(S) &= \mathcal{D} \\ S &\subseteq \mathcal{T} \end{aligned}$$

where $\text{cover}(S) = \{d \in \mathcal{D} : \gamma_d(d, s) \leq r \text{ for some } s \in S\}$ according to Definition 2.5, and γ_d is the norm or gauge associated with demand point $d \in \mathcal{D}$. We first discuss the feasibility of (CSL). Recall that $\mathcal{T}(d)$ is the set of all points on $\mathcal{T}$ which can be used to cover d, i.e., $\mathcal{T}(d) = \{s \in \mathcal{T} : \gamma_d(d, s) \leq r\}$.

Lemma 3.1. *Given an instance of (CSL), the following properties hold.*

1. *(CSL) has a feasible solution if and only if $\mathcal{T}(d) \neq \emptyset$ for all $d \in \mathcal{D}$.*
2. *If (CSL) has a feasible solution, then it also has a finite solution with cardinality less than or equal to $|\mathcal{D}|$.*

Proof. Part 1 is a direct consequence of Lemma 2.9. For the second part, if (CSL) has a feasible solution, we know that $\mathcal{T}(d) \neq \emptyset$ for all $d \in \mathcal{D}$. Choosing $s_d \in \mathcal{T}(d)$ for all $d \in \mathcal{D}$ yields a finite feasible solution

$$S = \{s_d : d \in \mathcal{D}\} \subseteq \mathcal{T}$$

with $|S| \leq |\mathcal{D}| < \infty$. □

Next, we discuss the complexity status of (CSL). It turns out that even the unweighted version of (CSL) is NP-hard in the special case that γ_d is the Euclidean distance for all $d \in \mathcal{D}$.

Theorem 3.2. *(CSL) is NP-hard.*

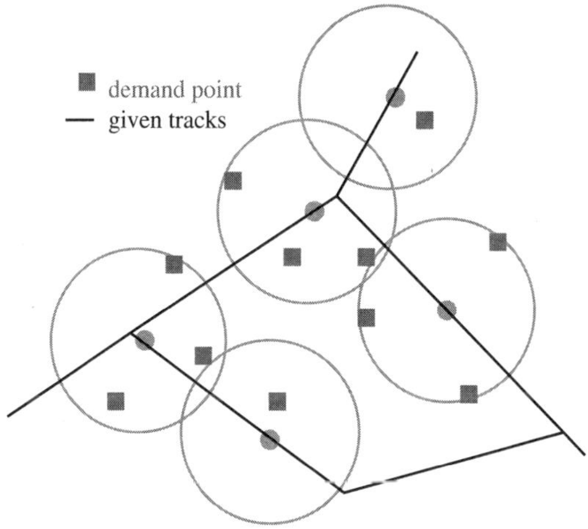

Fig. 3.1. A feasible solution for the Euclidean (CSL).

Proof. We show the NP-hardness of the special case of (CSL), in which we do not use weights (i.e., all traffic loads are one) and take the Euclidean distance as distance measure for all given demand points. I.e., the goal in this case is to cover all demand points with a minimal number of stops. Since γ_d is the Euclidean distance this problem has a nice geometric interpretation: We aim to cover a set of given points in the plane by discs whose center points are restricted to be in $\mathcal{T}$. For an illustration, see Figure 3.1. Formally, we define:

(Euclidean unweighted CSL-decision version) Given a finite set $\mathcal{D} \subseteq \mathbb{R}^2$, a set of points $\mathcal{T} = \cup_{e \in E} e \subseteq \mathbb{R}^2$ of the embedding of a planar connected graph with edge set E, and a positive integer $K \leq |\mathcal{D}|$, does there exist a collection of K discs $\mathcal{C}$ with radius r and center points in $\mathcal{T}$, such that each $d \in \mathcal{D}$ lies in at least one of the discs $C \in \mathcal{C}$?

The proof is based on a reduction of *Geometric Covering by Discs* to the above decision version of (CSL). Geometric Covering by Discs has been shown to be NP-complete [Joh82] and can be stated as follows:

(Geometric Covering by Discs) Given a finite set $\mathcal{D}$ of points in the plane and positive integers r and $K \leq |\mathcal{D}|$, can the points of $\mathcal{D}$ be covered by at most K discs of radius r?

Now take an instance of Geometric Covering by Discs. To construct an instance of the unweighted (CSL) we

- use the same set $\mathcal{D}$ of points,
- the same numbers r and K and
- define the set of edges E of a connected planar graph as follows: For each unordered pair of points d_1 and d_2 from $\mathcal{D}$,

(i) add the line segment from d_1 to d_2 as an edge and also
(ii) add a sufficiently large piece of the bisector of d_1 and d_2 to E.

Claim: $\mathcal{D}$ can be covered by at most K discs of radius r, if and only if $\mathcal{D}$ can be covered by at most K discs of radius r which all have their center points in $\mathcal{T}$.

To see this, assume that $\mathcal{D}$ can be covered by some collection $\mathcal{C}$, consisting of at most K discs of radius r. Then, for each disc $C \in \mathcal{C}$:

Case 1: If C contains no points of $\mathcal{D}$, then disregard C.
Case 2: If C contains only one point d of $\mathcal{D}$, then replace C by the disc with center point d (d is in $\mathcal{T}$, since all line segments from d to any other point from $\mathcal{D}$ are in $\mathcal{T}$) and radius r.
Case 3: If C contains a set of points $A \subseteq \mathcal{D}$ with $|A| \geq 2$ (i.e., C contains more than one point of $\mathcal{D}$), then replace C by a disc with center point q and radius r, where q is the center point of the smallest enclosing circle of A.
Since C covers A, the radius of the smallest enclosing circle is smaller than or equal to r, and hence the disc with radius r and center point q also covers A. Note that finding q is a well-known problem of location theory and can be done in linear time [Meg83].
Moreover it is known that q always lies on at least one bisector of points in A, see [EH72], such that q satisfies $q \in \mathcal{T}$.

In summary, $\mathcal{D}$ can be covered by at most K discs of radius r, all with center points in $\mathcal{T}$, if and only if it can be covered by at most K discs of radius K. This completes the proof. □

As a consequence, the unweighted and the weighted optimization versions of (CSL) are NP-hard problems. Nevertheless we will present efficient solution approaches which can be applied for a large class of special cases.

3.2 A Finite Dominating Set

The goal of this section is to reduce (CSL) to a set covering problem by determining a finite dominating set FDS, i.e., a finite set of candidates $\mathcal{S} \subseteq \mathcal{T}$, for which we know that it contains at least one optimal solution S^*, if the problem is feasible. Throughout this section, let us hence assume that (CSL) has a feasible solution, which is established easily by checking if

$$B_d^r \cap \mathcal{T} \neq \emptyset \text{ for all } d \in \mathcal{D}$$

(see Lemma 3.1). We define

$$\tilde{\mathcal{S}} = \{s \in \mathcal{T} : \gamma_d(d, s) = r \text{ for some } d \in \mathcal{D}\}, \tag{3.1}$$

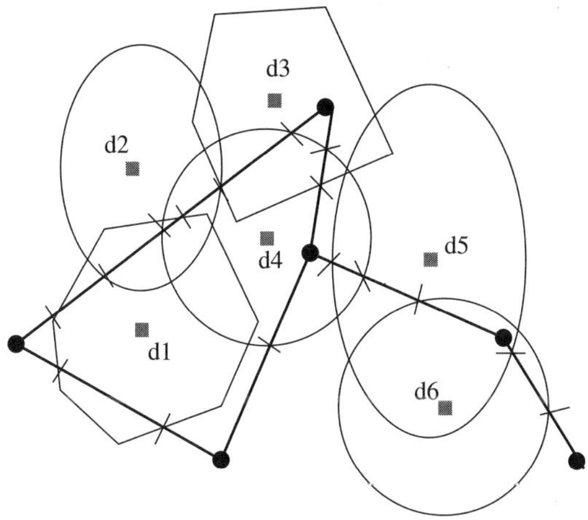

Fig. 3.2. The set of candidates along the tracks.

i.e., $\tilde{S}$ is given by the intersection points of $\mathcal{D}$ with the boundaries of the balls B_d^r, which are the unit balls B_d of radius r centered at the demand points d (see Figure 3.2).
To analyze $\tilde{S}$ we consider the situation along each edge $e \in E$ separately. For an edge $e = \{v_1^e, v_2^e\} \in E$ with endpoints $v_1^e, v_2^e \in V$, the definition $v_1^e \leq_e v_2^e$ naturally induces a (total) order for all points $s \in e$. This means we can talk about intervals as follows.

Notation 3.3.

- *Let $\leq_e$ denote the order along the line segment e, induced by $v_1^e \leq_e v_2^e$.*
- *For $s_1, s_2 \in e$ let $[s_1, s_2]_e = \{s \in e : s_1 \leq_e s \leq_e s_2\}$ denote the set of points on e between s_1 and s_2, and $]s_1, s_2[_e = \{s \in e : s_1 <_e s <_e s_2\}$ denote the set of points strictly between s_1 and s_2.*

Consequently, $\tilde{S}$ can be ordered along each edge $e \in E$. Unfortunately, it may happen that $|\tilde{S} \cap e| = \infty$, i.e., the boundary of some B_d^r contains a linear piece which coincides with e in infinitely many points, see $p_2 + rB_2$ in Figure 3.3. To overcome this problem, we first need the following simple lemma.

Lemma 3.4. *Let $e \in E$ be an edge and assume that $\mathcal{T} = e$ consists only of this edge. Then $\mathcal{T}(d)$ is a convex set for all $d \in \mathcal{D}$.*

Proof. Since $\mathcal{T} = e$ we know that $\mathcal{T}$ is a convex set in this case. According to Lemma 2.8, $\mathcal{T}(d) = \mathcal{T} \cap B_d^r$, i.e., $\mathcal{T}(d)$ is the intersection of two convex sets and thus is itself convex. □

Hence, for *all* $d \in \mathcal{D}$ either $B_d^r \cap e = \emptyset$, or

$$\mathcal{T}(d) \cap e = B_d^r \cap e = [f_d^e, l_d^e]_e \text{ for some points} f_d^e, l_d^e \in e.$$

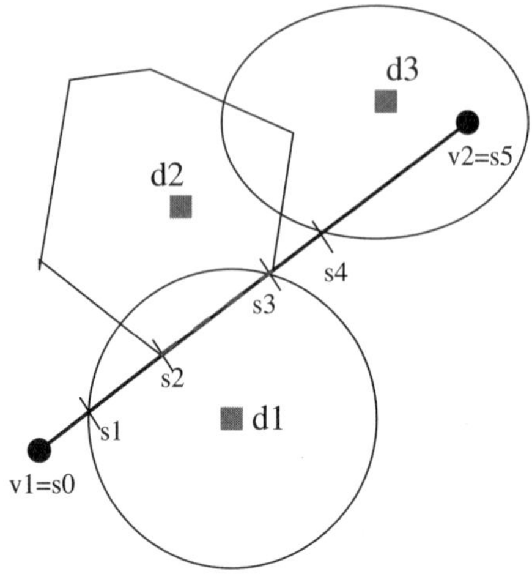

Fig. 3.3. The set of candidates on one edge $e \in E$. Between s_2 and s_3 all points are in $\tilde{\mathcal{S}}$, but not in $\mathcal{S}^e$.

We call f_d^e the *first* point of $\mathcal{T}(d)$ on edge e and l_d^e the *last* point of $\mathcal{T}(d)$, respectively. In the following, we will show that only these points f_d^e and l_d^e of the interval $\mathcal{T}(d)$ need to be considered, also in the case of $|\tilde{\mathcal{S}} \cap e| = \infty$. We define

Notation 3.5.

$$\mathcal{S}^e = \{v_1^e, v_2^e\} \cup \begin{cases} \bigcup_{d \in \mathcal{D}} \{f_d^e, l_d^e\} & \text{if } \bigcup_{d \in \mathcal{D}} \{f_d^e, l_d^e\} \text{ contains at least one} \\ & \text{point of the interior of } e \\ \{\frac{v_1^e + v_2^e}{2}\} & \text{otherwise.} \end{cases}$$

Note that instead of $\frac{v_1^e + v_2^e}{2}$ any other point in the interior of the edge e may be used. This case makes sure that at least one point of edge e is included in $\mathcal{S}^e$ covering the (unlikely) case that the whole edge e lies in the interior of at least one B_d^r, does not intersect any of the other $B_{d'}^r$, and satisfies $c_e < \min\{c_{v_1^e}, c_{v_2^e}\}$, i.e., has smaller costs than both of its endpoints.
If $c_{v_1^e} \le c_e$, or $c_{v_2^e} \le c_e$ it may happen that one of the endpoints of e is included in the optimal solution, hence both v_1^e and v_2^e must be in $\mathcal{S}^e$.
The result is a finite set of points

$$\mathcal{S}^e = \{s_0, s_1, \ldots, s_{N_e+1}\}$$

for which we assume $s_0 = v_1^e \le_e s_1 \cdots \le_e s_{N_e} \le s_{N_e+1} = v_2^e$. (The situation is depicted in Figure 3.3.)

Lemma 3.6. *Let e be an edge of E, and let $s \in]s_j, s_{j+1}[_e$ for some $j \in \{0, 1, \ldots, N_e\}$. Then*

$$\text{cover}(s) \subseteq \text{cover}(s_j) \cap \text{cover}(s_{j+1}).$$

Proof. Suppose $\text{cover}(s) \not\subseteq \text{cover}(s_j)$, i.e., there exists $d \in \mathcal{D}$ such that $\gamma_d(d,s) \leq r$ and $\gamma_d(d,s_j) > r$. Note that for any gauge γ_d the distance $\gamma_d(d,s)$ from d to s is continuous if we fix d and move s along a line. Hence, the intermediate value theorem yields a point $\tilde{s} \in]s_j, s]_e$ with $\gamma_d(d,\tilde{s}) = r$. Take $\tilde{s}$ minimal with respect to $<_e$ with this property, i.e., $\tilde{s} = f_d^e$ is the first point of $\mathcal{T}(d)$, and hence $\tilde{s} \in \mathcal{S}^e$. This contradicts the definition of s_j and s_{j+1} as consecutive points in $\mathcal{S}^e$. □

Now we are able to prove that

$$\mathcal{S} = \bigcup_{e \in E} \mathcal{S}^e \tag{3.2}$$

is, indeed, a finite dominating set.

Theorem 3.7. *Either (CSL) is infeasible, or there exists an optimal solution $S^* \subseteq \mathcal{S}$.*

Proof. Let $S^* \subseteq \mathcal{T}$ be optimal, but $S^* \not\subseteq \mathcal{S}$. The goal is to replace each $s \in S^* \setminus \mathcal{S}$ by a point in $\mathcal{S}$ without loosing feasibility or optimality. To this end, take some $s \in S^* \setminus \mathcal{S}$. Note that $s \notin V$, i.e., $g(s) = e \in E$, especially let

$$s \in]s_j, s_{j+1}[_e \text{ for some } j \in \{0, 1, \dots, N_e\}.$$

Since $\mathcal{S}$ contains at least one point of the interior of e, we can assume without loss of generality that $s_j \notin V$. For s_j, Lemma 3.6 yields that $\text{cover}(s) \subseteq \text{cover}(s_j)$ such that

$$S' = S^* \setminus \{s\} \cup \{s_j\}$$

still covers $\mathcal{D}$. Moreover, since $g(s) = g(s_j) = e$ both points have the same costs c_e such that we obtain $f_{\text{time}}(S') \leq f_{\text{time}}(S^*)$, which completes the proof. □

In the following special cases we do not need to investigate the complete set $\mathcal{S}$:

- The unweighted (CSL) either is infeasible, or there exists an optimal solution
$$S^* \subseteq \bigcup_{d \in \mathcal{D}, e \in E} \{f_d^e, l_d^e\}$$
if $\bigcup_{d \in \mathcal{D}, e \in E} \{f_d^e, l_d^e\} \neq \emptyset$, i.e., neither V nor the points $\frac{v_1+v_2}{2}$ are needed as candidates in this case.
- If γ_d is a strictly convex norm for all $d \in \mathcal{D}$ (e.g., if all γ_d are the Euclidean distance), we have that the candidate set $\tilde{\mathcal{S}}$ defined in (3.1) on page 24 already is a finite set, i.e.,
$$\tilde{\mathcal{S}} = \bigcup_{d \in \mathcal{D}, e \in E} \{f_d^e, l_d^e\} \subseteq \mathcal{S}.$$

- Together, in the unweighted Euclidean case (and for any other strictly convex norm), $\tilde{\mathcal{S}}$ suffices as finite dominating set, if $\tilde{\mathcal{S}} \neq \emptyset$. (Note that even this case is NP-hard.)

Note that by Theorem 3.7 we have transformed (CSL) into a *Location Set Covering Problem* , introduced originally in [TSRB71, TR73]. In our case, $\mathcal{S}$ is the discrete set of possible locations. Hence, (CSL) can be formulated as a set covering problem as follows: For all $s \in \mathcal{S}$ let x_s be a variable with the following meaning:

$$x_s = \begin{cases} 1 & \text{if } s \text{ is contained in the optimal solution} \\ 0 & \text{otherwise} \end{cases} .$$

Notation 3.8. *Define* $A^{\text{cov}} = (a_{ds})_{d \in \mathcal{D}, s \in \mathcal{S}}$ *with elements*

$$a_{ds} = \begin{cases} 1 & \text{if } \gamma_d(d,s) \leq r \\ 0 & \text{otherwise} \end{cases}$$

as the matrix containing the covering information. A^{cov} *is called the* **covering matrix**.

Then (CSL) is equivalent to the following set covering problem, where each row of the covering matrix A^{cov} corresponds to a demand point, while the columns represent the possible candidates.

$$\begin{aligned} \min \quad & cx \\ s.t.\ A^{\text{cov}} x &\geq \mathbf{1}_{|\mathcal{D}|} \\ x &\in \{0,1\}^{|\mathcal{S}|}, \end{aligned} \tag{3.3}$$

where $\mathbf{1} \in \mathbb{R}^{|\mathcal{D}|}$ is the vector consisting of a 1 in each component. The cost vector $c \in \mathbb{R}^{|\mathcal{S}|}$ in (3.3) is given by

$$c_s = c_{g(s)} \text{ for all } s \in \mathcal{S}.$$

As a consequence, to solve (CSL) any approach for set covering can be used. Still, the set covering problem is NP-hard. In the next sections we will develop a more efficient approach by taking advantage of the special structure of the covering matrix A^{cov}. We conclude Section 3.2 by summarizing the meaning of the elements a_{ds} of A^{cov}.

Lemma 3.9. *Let* A^{cov} *be the covering matrix of (CSL) as defined in Notation 3.8. Then*

$$a_{ds} = 1 \Longleftrightarrow \gamma_d(d,s) \leq r \Longleftrightarrow s \in \mathcal{T}(d) \Longleftrightarrow d \in \text{cover}(s).$$

3.3 Complete Cover Along a Polygonal Line

We start by investigating (CSL) along one single edge $e \subseteq \mathcal{T}$ and show that in this case, the problem can be solved efficiently. This is due to Lemma 3.4 (see page 25), which has a nice consequence for the structure of the coefficient matrix A^{cov} of the corresponding integer program (3.3), namely the matrix has the *consecutive ones property*, defined below.

Definition 3.10 (e.g., [GJ79a, NW88]). *Let A be a $(0,1)$-matrix.*

1. *A has the* **consecutive ones property (c1p)** *if for all rows i of A the following holds.*

$$a_{ik} = 1, a_{il} = 1, \text{ and } k \leq l \Longrightarrow a_{ij} = 1 \text{ for all } k \leq j \leq l.$$

2. *A is an* **interval matrix**, *if its transposed A^T has the consecutive ones property.*

Lemma 3.11. *Consider (CSL) in the special case that $\mathcal{T} = e$ consists of a single edge. Then there exists an order of $\mathcal{S}$ such that the resulting covering matrix A^{cov} of (CSL) has the consecutive ones property.*

Proof. Each column of A^{cov} represents a candidate $s \in \mathcal{S}$. Order the columns of A^{cov} according to the order of the candidates induced by $\leq_e$ (see Notation 3.3 on page 25). We show that according to this order, A^{cov} has the consecutive ones property. To this end, take a row of A^{cov} (belonging to a demand point $d \in \mathcal{D}$) such that

$$a_{ds_k} = 1 \text{ and } a_{ds_l} = 1 \text{ for some } s_k \leq_e s_l.$$

According to Lemma 3.9 this means $s_k \in \mathcal{T}(d)$ and $s_l \in \mathcal{T}(d)$. Since $\mathcal{T}(d)$ is convex (see Lemma 3.4) the whole segment $[s_k, s_l]$ lies in $\mathcal{T}(d)$. I.e., all s with $s_k \leq_e s \leq_e s_l$ satisfy $s \in \mathcal{T}(d)$, and hence $a_{ds} = 1$. □

In the next section we will develop various approaches for solving set covering problems efficiently in the case that the coefficient matrix has the consecutive ones property. But first we extend Lemma 3.11 to polygonal lines instead of one single edge. We use the following notation.

Notation 3.12.

- *If the graph $G = (V, E)$ is a simple path, its corresponding set of tracks $\mathcal{T}$ will be called a* **polygonal line**.
- *Let v_a,v_b be the endpoints of a polygonal line $\mathcal{T}$. Then let $\leq_{\mathcal{T}}$ denote the (total) order on $\mathcal{T}$ defined by $v_a \leq_{\mathcal{T}} v_b$.*
- *$I \subseteq \mathcal{T}$ is called an* **interval** *of the polygonal line $\mathcal{T}$ if there exist $s_1, s_2 \in \mathcal{T}$ such that $I = \{s \in e : s_1 \leq_{\mathcal{T}} s \leq_{\mathcal{T}} s_2\}$.*

Note that a subset I of a polygonal line $\mathcal{T}$ is an interval of $\mathcal{T}$ if and only if I is a connected part of $\mathcal{T}$.

Theorem 3.13. *Let $\mathcal{T}$ be a polygonal line. If $\mathcal{T}(d)$ is an interval of $\mathcal{T}$ for all $d \in \mathcal{D}$, then there exists an order of $\mathcal{S}$ such that the coefficient matrix of (CSL) has the consecutive ones property.*

Proof. Let v_a, v_b be the endpoints of the polygonal line $\mathcal{T}$. Since all candidates $\mathcal{S}$ are contained in $\mathcal{T}$ we can take the order induced by $v_a \leq_{\mathcal{T}} v_b$ and order the columns of A^{cov} with respect to $\leq_{\mathcal{T}}$. To show that according to this order, A^{cov} has the consecutive ones property, take a row d of A^{cov} and let

$$a_{ds_k} = 1 \text{ and } a_{ds_l} = 1 \text{ for some } s_k \leq_{\mathcal{T}} s_l.$$

According to Lemma 3.9 this means $s_k, s_l \in \mathcal{T}(d)$. Due to the assumptions of the theorem, $\mathcal{T}(d)$ is an interval included in $\mathcal{T}$, i.e., there exist two points $f_d, l_d \in \mathcal{T}$, such that

$$\mathcal{T}(d) = \{s \in \mathcal{T} : f_d \leq_{\mathcal{T}} s \leq_{\mathcal{T}} l_d\}.$$

But this means that all s with $s_k \leq_{\mathcal{T}} s \leq_{\mathcal{T}} s_l$ satisfy $s \in \mathcal{T}(d)$, and hence $a_{ds} = 1$. □

Note that Lemma 3.11 is a special case of Theorem 3.13. Geometrically, the conditions of the theorem are satisfied, if

$$\text{crit}(e_1, e_2) = (\text{cover}(e_1) \cap \text{cover}(e_2)) \setminus \text{cover}(e_1 \cap e_2) = \emptyset$$

for all pairs of edges $e_1, e_2 \in E$.

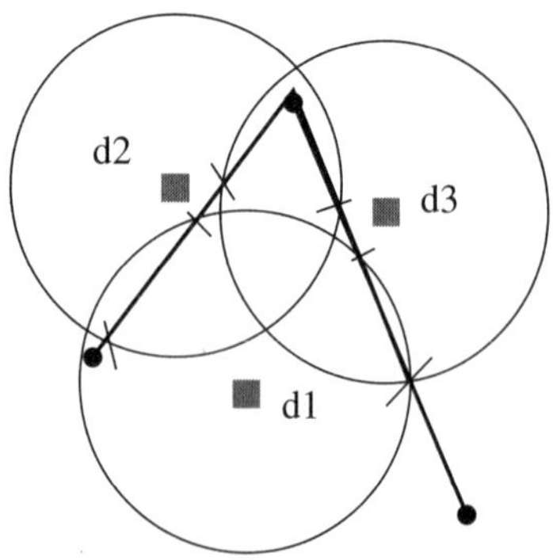

Fig. 3.4. An instance of (CSL) on a polygonal line without consecutive ones property.

An example of a polygonal line not satisfying the condition of Theorem 3.13 with a coefficient matrix without consecutive ones property is given in Figure 3.4. In this example, G is a simple path consisting of three nodes. Numbering the candidates in $\tilde{\mathcal{S}}$ (which is sufficient in the unweighted Euclidean case, see the third special case on page 28) from left to right, A^{cov} is given by

$$A^{\text{cov}} = \begin{pmatrix} 1\,1\,0\,0\,1\,1 \\ 1\,1\,1\,1\,0\,0 \\ 0\,0\,1\,1\,1\,1 \end{pmatrix},$$

which cannot be reordered to satisfy the consecutive ones property.

On the other hand, Figure 3.5 shows an example of a polygonal line together with a set of demand points $\mathcal{D}$ satisfying the consecutive ones property. The reason why it is advantageous that the covering matrix of a given instance of (CSL) satisfies the consecutive ones property becomes clear in the following result.

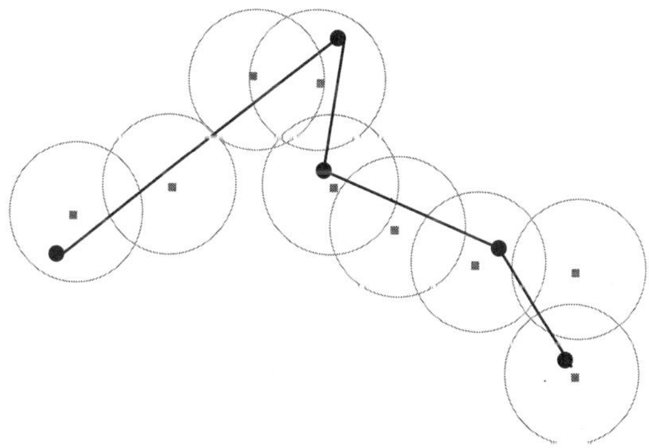

Fig. 3.5. An instance of (CSL) on a polygonal line with consecutive ones property.

Theorem 3.14. *(CSL) can be solved in polynomial time by linear programming if the conditions of Theorem 3.13 are satisfied. In particular, (CSL) can be solved by linear programming in the special case that $\mathcal{T} = e$ is a single edge.*

Proof. From Theorem 3.13 we know that the coefficient matrix A^{cov} of the integer programming formulation (3.3) has the consecutive ones property. This means, A^{cov} is a totally unimodular matrix (see, e.g., Corollary 2.10 in Chapter III.1.1 of [NW88]) and hence the IP-relaxation of (3.3) solves the integer program, see Appendix A. □

Other efficient procedures for solving set covering problems where the covering matrix has the consecutive ones property are presented in the next section.

3.4 Set Covering With Consecutive Ones Property

The problem we consider in this section is a set covering problem with consecutive ones property, which we will denote by (SCP-c1p). Although we use the notation already introduced for the stop location problem, we remark that the methods presented in this section are applicable to any set covering problem with consecutive ones property. Let us consider

(SCP-c1p)

$$\begin{aligned} \min \quad & cx \\ \text{s.t. } & A^{\text{cov}} x \geq \mathbf{1}_{|\mathcal{D}|} \\ & x \in \{0,1\}^{|\mathcal{S}|}, \end{aligned}$$

in which we assume that the covering matrix A^{cov} has the consecutive ones property, and $\mathbf{1}_{|\mathcal{D}|} \in \mathbb{R}^{|\mathcal{D}|}$ is a vector with a 1 in each component.
In the stop location problem, the set of rows $\mathcal{D} = \{1, \ldots, |\mathcal{D}|\}$ corresponds to the demand points, while the set of columns $\mathcal{S} = \{1, \ldots, |\mathcal{S}|\}$ corresponds to the candidates of the finite dominating set, ordered in such a way that the resulting coefficient matrix has the consecutive ones property. Finally, $c = (c_s)_{s \in \mathcal{S}}$ is defined by the costs $c_s = c_{g(s)}$ of the new stops. In our application we usually have $|\mathcal{D}| \leq |\mathcal{S}|$. However, in the special case of Lemma 3.11 we know that no more than two candidates can arise from one demand point, such that $O(|\mathcal{D}|)=O(|\mathcal{S}|)$. Similarly, in the case of Theorem 3.13 we obtain $O(|\mathcal{S}|)=O(|\mathcal{D}|) + O(|V|)$.

As already stated in Theorem 3.14, matrices with consecutive ones property are totally unimodular and (SCP-c1p) can hence be solved by linear programming.

In a more efficient approach we use the fact that the transpose of a matrix with consecutive ones property is an interval matrix, and hence a network matrix. Since there exists an optimal solution satisfying $x_s \leq 1$ for all $s \in \mathcal{S}$, we omit these constraints and obtain the following linear program as the dual of the set covering problem.

(Dual-SCP)

$$\begin{aligned} \max \quad & \mathbf{1}\eta \\ s.t. \ & (A^{\text{cov}})^T \eta \leq c \\ & \eta \geq 0. \end{aligned}$$

Note that, since A^{cov} is totally unimodular, the optimal solution values of (SCP-c1p) and its dual formulation (Dual-SCP) are equal.

Following the approach of Example 3.2. in Chapter III.1.3 of [NW88], this dual formulation can be reformulated as a network flow problem in an acyclic network. This network is constructed by interpreting the rows of $(A^{\text{cov}})^T$ as arcs and the columns as paths. One starts by defining the set of nodes as

$$V_{\text{flow1}} = \{0, 1, \ldots, |\mathcal{S}|\},$$

and by constructing an arc $(s-1, s) \in E_{\text{flow1}}$ for each row s of $(A^{\text{cov}})^T$. Furthermore, each column d of $(A^{\text{cov}})^T$ can be interpreted as a path which is composed of the edges $(s-1, s)$ with $a_{ds} = 1$. For such a path we add one additional arc to E_{flow1}, namely the one replacing the respective path. These arcs correspond to the dual variables η_d. Since all entries in A^{cov} are positive, the network is acyclic. Defining $d_0 = c_1$, $d_s = c_{s+1} - c_s$ for $s = 1, \ldots, |\mathcal{S}| - 1$, and $d_{|\mathcal{S}|} = -c_{|\mathcal{S}|}$ as the demand of the respective node in V_{flow1}, and setting 0 as the cost of arc $(s-1, s)$, and 1 as the costs for all other arcs, one finally obtains an equivalent *min-cost flow problem* in an acyclic digraph with $|\mathcal{S}|+1$ nodes, see [NW88] for more details.

In this section, however, we propose a new approach for solving set covering problems with consecutive ones property. This approach transforms the set covering problem into a *shortest path problem* in a directed acyclic network with $|\mathcal{S}| + 2$ nodes.

Taking the given order of the columns of A^{cov} as fixed, we can talk about $s_1 < s_2$ or $\min S$ for $S \subseteq \mathcal{S}$, recalling that in the stop location problem this refers to the order $\leq_{\mathcal{T}}$. We need the following notation.

Notation 3.15. *Let A be a matrix containing no zero rows. For $s \in \mathcal{S}$ and $d \in \mathcal{D}$ define*

$$\begin{aligned}
\text{cover}(s) &= \{d \in \mathcal{D} : a_{ds} = 1\} \\
T(d) &= \{s \in \mathcal{S} : a_{ds} = 1\} \\
f_d &= \min\{s \in \mathcal{S} : a_{ds} = 1\} \text{ is the first element of } T(d) \\
l_d &= \max\{s \in \mathcal{S} : a_{ds} = 1\} \text{ is the last element of } T(d).
\end{aligned}$$

Then, if A^{cov} has the consecutive ones property, we conclude that for all $d \in \mathcal{D}$,

$$T(d) = \{s \in \mathcal{S} : f_d \leq s \leq l_d\}.$$

Note that for the stop location problem,

$$\begin{aligned}
T(d) &= \mathcal{T}(d) \cap \mathcal{S} = \{s \in \mathcal{S} : \gamma_d(d, s) \leq r\} \text{ and} \\
\text{cover}(s) &= \{d \in \mathcal{D} : \gamma_d(d, s) \leq r\}
\end{aligned}$$

have been used before, but since we do not want to use any properties of (CSL) in this section we redefined both sets using only the covering matrix A^{cov}.

Without loss of generality let us assume that A^{cov} does not contain any zero column, since such columns do not cover any row and hence will never appear in an optimal solution. Other reduction possibilities which were proposed in [TR73] are listed below in the notation of the stop location problem. Note that these rules can be applied to any set covering problem (SCP), with or without consecutive ones property.

Lemma 3.16 ([TR73]).

1. *Problem (SCP) has a feasible solution if and only if* $T(d) \neq \emptyset$ *for all* $d \in \mathcal{D}$.
2. *If* $T(d_1) \subseteq T(d_2)$, *an optimal solution of problem (SCP) can be found by considering the reduced problem without row* d_2.
3. *If* $\text{cover}(s_1) \subseteq \text{cover}(s_2)$ *and* $c_{s_1} \geq c_{s_2}$ *then there exists an optimal solution of problem (SCP) with* $x_{s_1} = 0$, *i.e., it is sufficient to consider the reduced problem without column* s_1.
4. *If* $T(d) = \{s\}$ *then in all optimal solutions* $x_s = 1$ *holds, and it is sufficient to consider the reduced problem without column* s *and without all rows* $d' \in \text{cover}(s)$.

We first show that A^{cov} can be transformed into the following more convenient form.

Definition 3.17. *A matrix* A^{cov} *with consecutive ones property and without zero rows is called* **monotone** *if* $f_1 \leq f_2 \cdots \leq f_{|\mathcal{D}|}$ *and* $l_1 \leq l_2 \cdots \leq l_{|\mathcal{D}|}$ *hold simultaneously. Furthermore, if* $f_1 < f_2 \cdots < f_{|\mathcal{D}|}$ *and* $l_1 < l_2 \cdots < l_{|\mathcal{D}|}$, A^{cov} *is called* **strictly monotone**.

Lemma 3.18. *Let* A^{cov}, c *be the input data of a (feasible) set covering problem (SCP-c1p) with consecutive ones property. Then there exists an equivalent set covering problem with input data* $A^{\text{cov}}_{\text{mon}}, c$ *such that* $A^{\text{cov}}_{\text{mon}}$ *is a strictly monotone matrix, possibly with fewer rows than* A^{cov}.

Proof. The proof works by first sorting the rows of A^{cov} according to f_d and then applying the second reduction rule of Lemma 3.16 to eliminate rows until strictly monotonicity is obtained. □

This can be performed efficiently as follows:

Algorithm 2: Transforming a matrix with consecutive ones property into strictly monotone form

---

Input: Matrix A^{cov} with $|\mathcal{D}|$ (non-zero) rows and consecutive ones property.
Output: Strictly monotone matrix.
Step 1. Order the rows of A^{cov} such that $f_1 \leq f_2 \leq \ldots \leq f_{|\mathcal{D}|}$. Set $D = |\mathcal{D}|$.
Step 2. If $f_1 < f_2 \ldots < f_D$ set $d^0 = 1$ and goto 4. Otherwise choose d, d' such that $f_d = f_{d'}$.
Step 3. (Reduction 1) If $l_d \geq l_{m'}$: delete row d, otherwise delete row d'. Let $D = D - 1$, and rename $f_i, i = 1, \ldots, D$. Goto 2.
Step 4. $d = \text{argmin}\{l_{d'} : d' \geq d^0\}$. If the minimum is not unique, choose the one with the larger row index d.
Step 5. (Reduction 2) Delete all rows d' with $d^0 \leq d' < d$.
Step 6. If $d \geq D - 1$ stop, otherwise set $d^0 = d + 1$ and return to 4.

The Unweighted Set Covering Problem With Consecutive Ones Property

A special case is the *unweighted set covering problem* in which all $c_s = 1$. In this case the goal is to cover all rows with a minimal number of columns of A^{cov}. Since all $c_s = 1$ we can not only apply part 2, but also part 3 of Lemma 3.16, reducing the number of columns of A^{cov}. It turns out that a matrix A^{cov} with consecutive ones property can be reduced to a (smaller) unit matrix of an equivalent set covering problem. Using part 4 of Lemma 3.16 this (smaller) set covering problem can then be solved easily. Here, we propose the following approach for solving (SCP-c1p) in the unweighted case.

Algorithm 3: Solving the unweighted set covering problem with consecutive ones property

---

`Input:` Matrix A^{cov} with the consecutive ones property.
`Output:` An optimal solution S^* of (SCP).
`Step 1.` If A^{cov} contains no zero row, use Algorithm 2 to transform A^{cov} into a strictly monotone matrix, set $d = 1$, $S^* = \emptyset$.
Otherwise stop: Problem infeasible.
`Step 2.` $S^* = S^* \cup \{l_d\}$
`Step 3.` If $\{d' : f_{d'} > l_d\} \neq \emptyset$ choose $d = \min\{d' : f_{d'} > l_d\}$ and goto 2, otherwise stop. Output: S^*

Theorem 3.19. *Algorithm 3 finds an optimal solution of the unweighted set covering problem.*

Proof. Let $S = \{s_1, \ldots, s_p\}$ be the output of Algorithm 3. To each column s in S belongs a row $\bar{d}$ such that $l_{\bar{d}} = s$ (step 2). Let $\bar{\mathcal{D}} = \{\bar{d_1}, \bar{d_2}, \ldots, \bar{d_p}\}$. Then $\bar{d_1} < \bar{d_2} < \cdots \bar{d_p}$, since for $d_i \geq d_{i+1}$ the monotonicity of A^{cov} would imply $f_{d_i} \geq f_{d_{i+1}}$ which is a contradiction to

$$f_{d_{i+1}} > l_{d_i} \geq f_{d_i}, \text{ see step 3.}$$

Moreover, $\bar{d_1} = 1$ (step 1).

Feasibility: We show that $l_{\bar{d_j}}$ covers all rows d with $\bar{d_j} \leq d \leq \bar{d}_{j+1} - 1$: Note that $\bar{d}_{j+1} = \min\{d' : f_{d'} > l_{\bar{d_j}}\}$. Hence,

$$\begin{aligned} d < \bar{d}_{j+1} &\Longrightarrow f_d \leq l_{\bar{d_j}} \\ d \geq \bar{d_j} &\Longrightarrow l_d \geq l_{\bar{d_j}}. \end{aligned}$$

Together, $a_{d\, l_{\bar{d_j}}} = 1$, i.e., $l_{\bar{d_j}}$ covers d.

Optimality: Let $\tilde{S} = \{\tilde{s}_1, \ldots, \tilde{s_q}\}$ (ordered) be any solution of (SCP) with $q < p$. Since row $\bar{d}_1 = 1$ must be covered, we know that $\tilde{s}_1 \leq l_1 = s_1$. Also, $\bar{d}_2$ must be covered, and together with $f_{\bar{d}_2} > l_{\bar{d}_1}$ (step 3) this yields $\tilde{s}_2 \leq l_{\bar{d}_2} = s_2$. Iterating this argument, we finally get that $\tilde{s}_q \leq l_{\bar{d}_q} = s_q$, but consequently, the rows $\bar{d}_{q+1}, \ldots, |\mathcal{D}|$ are not covered by $\tilde{S}$, hence any smaller solution is infeasible. □

Note that we can also directly apply Algorithms 2 and 3 to the stop location problem on a single edge (or on a polygonal line, if the assumptions of Theorem 3.13 are satisfied and the costs of the new stops are all equal). The idea is to calculate the intervals $\mathcal{T}(d) = [f_d, l_d]$ for all demand points d, order them with respect to their endpoints l_d, and then to choose successively the last possible candidate for covering the first uncovered demand point. A reformulation in algorithmic form is given next. In Section 5.1 we will show how to extend this approach to demand regions instead of demand points.

Algorithm 4: Solving (CSL) along a polygonal line

Input: Polygonal line $\mathcal{T}$, satisfying the assumptions of Theorem 3.13, $\mathcal{D}$, γ_d for all $d \in \mathcal{D}$, and $c_e = 1$ for all edges $e \subseteq \mathcal{T}$, $c_v = 1$ for all nodes $v \in \mathcal{T}$.
Output: An optimal solution S^* of (CSL).
Step 1. Calculate $\mathcal{T}(d) = [f_d, l_d]$ for all $d \in \mathcal{D}$ and order $\mathcal{D} = \{d_1, \ldots, d_{|\mathcal{D}|}\}$ according to the endpoints l_d of $\mathcal{T}(d)$ with respect to $\leq_{\mathcal{T}}$. If $\mathcal{T}(d) = \emptyset$ stop: Problem infeasible. Otherwise set $d = 1$, $S^* = \emptyset$.
Step 2.

$$S^* = S^* \cup \{\min_{d \in \mathcal{D}} l_d\}$$
$$\mathcal{D} = \mathcal{D} \setminus \text{cover}(l_d)$$

Step 3. If $\{d' : f_{d'} > l_d\} \neq \emptyset$ goto 2, otherwise stop. Output: S^*

The Weighted Set Covering Problem With Consecutive Ones Property

We now come back to the weighted set covering problem (SCP-c1p) with consecutive ones property. Analogously to f_d, l_d we define

Notation 3.20.

$$\bar{f}_s = \min\{d \in \mathcal{D} : a_{ds} = 1\}$$
$$\bar{l}_s = \max\{d \in \mathcal{D} : a_{ds} = 1\}.$$

We need the following observations.

Lemma 3.21.

1. *Let A^{cov} be a monotone matrix (satisfying the consecutive ones property). Then A^{cov} is an interval matrix, i.e., $(A^{\text{cov}})^T$ also has the consecutive ones property. Moreover, $(A^{\text{cov}})^T$ is also monotone.*
2. *If $S = \{s_1, s_2, \ldots, s_p\} \subseteq \mathcal{S}$ is a cover of A^{cov} with $s_1 < s_2 < \cdots < s_p$ then $\bar{f}_{s_1} = 1$ and $\bar{l}_{s_p} = |\mathcal{D}|$.*

Proof.

1. Let A^{cov} be monotone. We first show that A^{cov} is an interval matrix. Assume the contrary, i.e., suppose that there exists a column s of A^{cov} and indices $d_1 < d_2 < d_3$ such that

$$\begin{aligned} a_{d_1 s} &= 1, \\ a_{d_2 s} &= 0, \\ a_{d_3 s} &= 1. \end{aligned}$$

   This yields $d_{d_1} \leq s \leq l_{d_1}$ and $d_{d_3} \leq s \leq l_{d_3}$, while for d_2 we obtain either $f_{d_2} > s$ or $l_{d_2} < s$. Both cases contradict the monotonicity of A^{cov}.

   For the monotonicity of the interval matrix we take $s_1 < s_2$. Assume $\bar{f}_{s_1} > \bar{f}_{s_2}$. Then $a_{\bar{f}_{s_2}, s_1} = 0$, yielding that

$$f_{\bar{f}_{s_2}} > f_{\bar{f}_{s_1}},$$

   a contradiction to the monotonicity of A^{cov}. The monotonicity of $\bar{l}_s$ is shown analogously.
2. For the second part of the lemma assume that $\bar{f}_{s_1} \neq 1$ in a cover S with smallest element s_1. Since A^{cov} has no zero columns we know that $a_{11} = 1$. Together with $\bar{f}_{s_1} > 1$ we conclude from the consecutive ones property of A^{cov} that $a_{1s} = 0$ for all $s \geq s_1$, in particular for all $s \in S$. This means that row 1 is not covered by S, a contradiction. Analogously, if $\bar{l}_{s_p} \neq |\mathcal{D}|$ we obtain $a_{|\mathcal{D}|s_p} = 0$ and thus $a_{|\mathcal{D}|s} = 0$ for all $s \leq s_p$, hence row $|\mathcal{D}|$ is not covered by S, a contradiction. □

Given a monotone matrix A^{cov} we are now in the position to define the following directed acyclic graph.

Notation 3.22. *The* **set covering digraph** $G_{SC} = (V_{SC}, E_{SC})$ *is defined by*

$$\begin{aligned} V_{SC} &= \mathcal{S} \cup \{s, t\} \quad \text{and} \\ E_{SC} &= \{(i,j) : i < j \text{ and } \bar{f}_j \leq \bar{l}_i + 1\} \cup \{(s,i) : \bar{f}_i = 1\} \cup \{(i,t) : \bar{l}_i = |\mathcal{D}|\}. \end{aligned}$$

Furthermore, for each edge (i,j) we associate a cost

$$c_{ij} = \begin{cases} c_j & \text{if } j \neq t \\ 0 & \text{if } j = t \end{cases}.$$

As an example, consider the matrix

$$A^{\text{cov}} = \begin{pmatrix} 1\,1\,1\,0\,0\,0 \\ 0\,1\,1\,1\,0\,0 \\ 0\,0\,1\,1\,1\,0 \\ 0\,0\,0\,0\,1\,1 \end{pmatrix}.$$

The digraph G_{SC} corresponding to A^{cov} hence has eight nodes, and is shown in Figure 3.6.

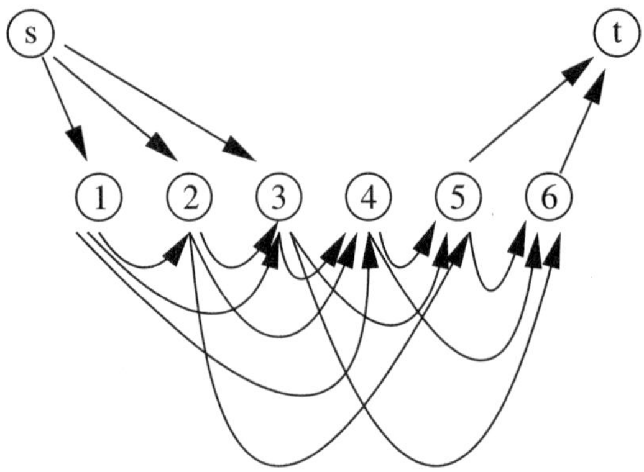

Fig. 3.6. The digraph G_{SC} for the example.

Since G_{SC} is acyclic, each node set $S \subseteq \mathcal{S}$ uniquely defines an s-t-path in G_{SC} by adding the nodes s and t to S. This justifies the notation of the next theorem.

Theorem 3.23. *Let $S \subseteq \mathcal{S}$. Then* $\text{cover}(S) = \mathcal{D}$ *if and only if $S \cup \{s, t\}$ is an s-t-path in G_{SC}.*

Proof. Let $S = \{s_1, s_2, \ldots, s_p\}$ with $s_1 < s_2 < \cdots < s_p$.

1. Let $S \cup \{s, t\}$ be an s-t-path in G_{SC}, and assume that S is not a cover. Choose an uncovered row d_0 with minimal index. Note that $d_0 \neq 1$ since $(s, s_1) \in E_{SC}$ meaning that $\bar{f}_{s_1} = 1$, i.e., $a_{1s_1} = 1$ and row 1 is covered. We hence can consider row $d_0 - 1$. Since d_0 is chosen minimal we know that row $d_0 - 1$ is covered, say by $s_i \in S$, and choose s_i with maximal index. Then $\bar{l}_{s_i} = d_0 - 1$. We distinguish two cases.
   $i < p$: Since $(s_i, s_{i+1}) \in E_{SC}$ we obtain that $\bar{f}_{s_{i+1}} \leq \bar{l}_{s_i} + 1 = d_0$. According to the monotonicity of $(A^{\text{cov}})^T$ we furthermore have $d_0 - 1 = \bar{l}_{s_i} \leq \bar{l}_{s_{i+1}}$. Together we conclude that either $d_0 - 1$ or d_0 is covered by S, a contradiction.
   $i = p$: Then $(s_i, t) \in E_{SC}$, such that $\bar{l}_{s_i} = |\mathcal{D}|$ yields $d_0 - 1 = |\mathcal{D}|$, a contradiction.

2. Now let S be a cover.
   - Then $\bar{f}_{s_1} = 1$ (Part 2 of Lemma 3.21) and hence $(s, s_1) \in E_{SC}$.
   - Analogously, $\bar{l}_{s_p} = |\mathcal{D}|$ yielding $(s_p, t) \in E_{SC}$.
   - Assume $(s_i, s_{i+1}) \notin E_{SC}$. Then $\bar{f}_{s_{i+1}} > \bar{l}_{s_i}+1$. Due to the monotonicity of $(A^{\text{cov}})^T$ we get:

$$\bar{f}_{s_j} \geq \bar{f}_{s_{i+1}} > \bar{l}_{s_i} + 1 \text{ for all } j \geq i+1, \text{ and}$$
$$\bar{l}_{s_j} \leq \bar{l}_{s_i} \text{ for all } j \leq i.$$

Together, row $\bar{l}_{s_i} + 1$ is not covered, a contradiction. □

Since the cost of a cover equals the cost of the corresponding path and vice versa, we finally get the following result.

Corollary 3.24. *A shortest s-t-path in G_{SC} represents a minimal cost cover and vice versa.*

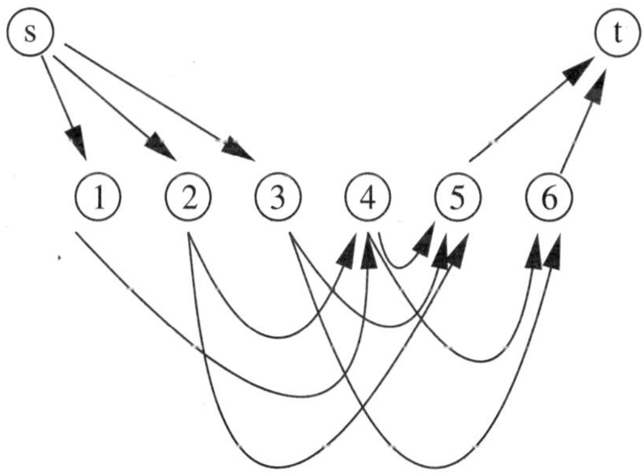

Fig. 3.7. The reduced digraph G'_{SC} for the example.

The corollary justifies the correctness of the next algorithm, which can further be improved by using only the reduced edge set

$$E'_{SC} = \{(i,j) : \bar{l}_i{+}1 \in \text{cover}(j)\} \cup \{(s,i) : 1 \in \text{cover}(i)\} \cup \{(i,t) : |\mathcal{D}| \in \text{cover}(i)\},$$

since it still contains all non-reducible covers, i.e., all covers S which satisfy that no $\tilde{S} \subset S$ also is a cover. For the matrix

$$A^{\text{cov}} = \begin{pmatrix} 1\,1\,1\,0\,0\,0 \\ 0\,1\,1\,1\,0\,0 \\ 0\,0\,1\,1\,1\,0 \\ 0\,0\,0\,0\,1\,1 \end{pmatrix}$$

two edges of E_{SC} need not be considered in $E_{SC'}$. Namely the edge from 3 to 4, since, if column 3 is chosen, row 3 is already covered and hence there is no need of choosing column 4. Analogously, the edge from 5 to 6 is omitted. Further reductions are possible, including the edges from 1 to 2, from 1 to 3, and from 2 to 3. The resulting digraph is depicted in Figure 3.7.

Algorithm 5: Solving the set covering problem with consecutive ones property

Input: Matrix A^{cov} with consecutive ones property, cost vector c.
Output: An optimal solution S^* of (SCP).
Step 1. If A^{cov} contains no zero row, use Algorithm 2 to transform A^{cov} into a strictly monotone matrix $A^{\text{cov}}_{\text{mon}}$. Otherwise stop: Problem infeasible.
Step 2. Derive the graph $G_{SC} = (V_{SC}, E_{SC})$ from $A^{\text{cov}}_{\text{mon}}$.
Step 3. Find a shortest s-t-path S in G_{SC} by a shortest path algorithm.
Output: $S^* = S \setminus \{s, t\}$.

Summarizing, we described the following two approaches.

- Using that $(A^{\text{cov}})^T$ is an interval matrix yields a network flow problem in an acyclic digraph with $|S| + 1$ nodes.
- Using the result of Theorem 3.23 yields a shortest path problem in an acyclic digraph with $|S| + 2$ nodes.

The numerical results obtained in the masters thesis of [Con02] indicate that solving set covering problems with consecutive ones property by an application of Algorithms 2 and 5 is much more efficient than solving the corresponding network flow problem. Note that other – very efficient – approaches for set covering problems with consecutive ones property using parametric shortest path techniques have recently been obtained by [HL05].

3.5 Complete Cover in a Realistic Network

We now turn our attention again to (CSL) in a general network. Unfortunately, nodes of the given graph G with degree higher than 2 can easily destroy the consecutive ones property, as Figure 3.8 shows. Although in this example $\mathcal{T}(d)$ is an interval for all three demand points d_1, d_2, and d_3, the coefficient matrix of (CSL) in this case is

$$A^{\text{cov}} = \begin{pmatrix} 1 & 1 & 0 & 1 \\ 1 & 1 & 1 & 0 \\ 1 & 0 & 1 & 1 \end{pmatrix},$$

which cannot be reordered to obtain the consecutive ones property. (Note that for the sake of simplicity in this example a very special situation is depicted, in which the unit balls B_d^r all pass through v_0 and through the same intersection points. But also more general instances do not result in a matrix with consecutive ones property.)

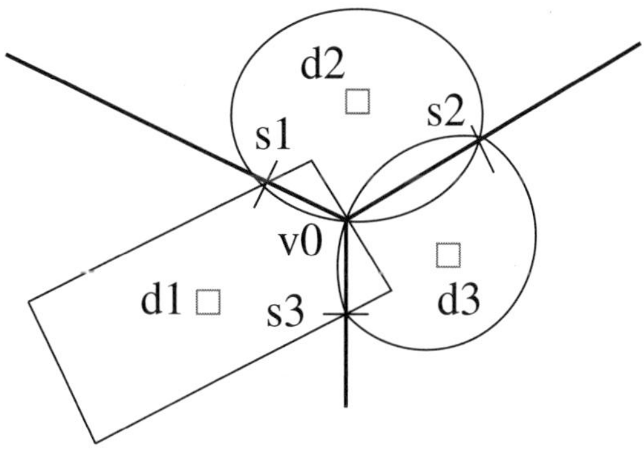

Fig. 3.8. An instance of (CSL) without consecutive ones property.

Nevertheless, we now will develop a decomposition result based on the following notation.

Notation 3.25. *Given* $G = (V, E)$ *and* $\mathcal{D}$ *the* **cover graph** $G_{\text{cover}} = (V_{\text{cover}}, E_{\text{cover}})$ *is a bipartite graph with*

- $V_{\text{cover}} = \mathcal{D} \cup E$, *and*
- $E_{\text{cover}} = \{\{d, e\} : \mathcal{T}(d) \cap e \neq \emptyset\}$.

The cover graph belonging to the example in Figure 3.9 is depicted in Figure 3.10. Note that this graph consists of four components:

$$\begin{array}{ll} \mathcal{D}_1 = \{d_1, d_2, d_3, d_4\}, & E_1 = \{e_1, e_2\} \\ \mathcal{D}_2 = \{d_5, d_6, d_7, d_8, d_9\}, & E_2 = \{e_3, e_4\} \\ \mathcal{D}_3 = \{d_{10}, d_{11}, d_{12}, d_{13}, d_{14}\}, & E_3 = \{e_5, e_6\} \\ \mathcal{D}_4 = \{d_{15}\}, & E_4 = \emptyset. \end{array}$$

These components can be used to decompose (CSL) as follows.

Notation 3.26. *Let* $\mathcal{D}_k \cup E_k$, $k = 1, \ldots, K$ *be the node sets of components of* G_{cover}, *i.e.,*

1. $\mathcal{D}_k \subseteq \mathcal{D}$, $E_k \subseteq E$ *for all* $k = 1, \ldots, K$,
2. $\bigcup_{k=1,\ldots,K} \mathcal{D}_k = \mathcal{D}$, $\bigcup_{k=1,\ldots,K} E_k = E$,
3. *if* $d \in \mathcal{D}_k$, $e \in E_l$, *and* $\{d, e\} \in E_{\text{cover}}$ *then* $k = l$.

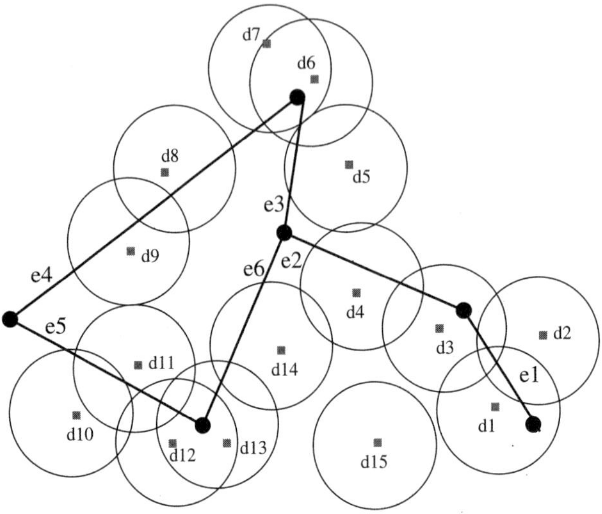

Fig. 3.9. Example graph to illustrate Notation 3.25 for the Euclidean distance.

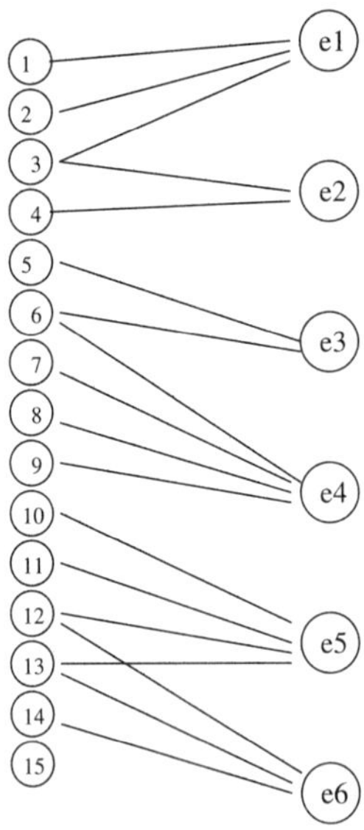

Fig. 3.10. The *cover graph* belonging to the graph in Figure 3.9.

Note that we do not require that the components are connected, but for solving the problem it is preferable to get a large number K of components. From the components of the cover graph we now define the following subproblems.

Notation 3.27. *Given $\mathcal{D}_k \cup E_k$, $k = 1, \ldots, K$ as components of G_{cover}, define subproblem (CSL(k)) for all $k = 1, \ldots, K$ by*

- *the reduced graph $G_k = (V(E_k), E_k)$, where $V(E_k)$ contains the nodes of the original graph G which are incident with at least one edge in E_k, and by*
- *the reduced set $\mathcal{D}_k$.*

Theorem 3.28. *Consider (CSL) in a graph G with demand points $\mathcal{D}$. Let (CSL(k)) for $k = 1, \ldots, K$ be subproblems defined by components of G_{cover} as specified in Notation 3.27. Then the following statements hold.*

1. *(CSL) is feasible if and only if (CSL(k)) is feasible for all $k = 1, \ldots, K$.*
2. *Let (CSL) be feasible and let S_k^* be an optimal solution for (CSL(k)), $k = 1, \ldots, K$. Then $\bigcup_{k=1,\ldots,K} S_k^*$ is an optimal solution for (CSL). I.e., each of the subproblems can be solved independently.*

Proof.

1. According to part 1 of Lemma 3.1 (CSL) is feasible if and only if $\mathcal{T}(d) \neq \emptyset$ for all $d \in \mathcal{D}$, which is equivalent to requiring that for each $d \in \mathcal{D}$ there exists an edge $\{d, e\} \in E_{\text{cover}}$.
   $\Longrightarrow$: Let (CSL) be feasible. Consider (CSL(k)). Then for each $d \in \mathcal{D}_k$ there exists $e \in E$ with $\{d, e\} \in E_{\text{cover}}$ and due to 3. in Notation 3.26 $e \in E_k$. Hence $\mathcal{T}(d) \cap E_k \neq \emptyset$ for all $d \in \mathcal{D}_k$, thus (CSL(k)) is feasible.
   $\Longleftarrow$: On the other hand, if (CSL) is not feasible, there exists $d \in \mathcal{D}$ which is an isolated point in G_{cover}. Let $d \in \mathcal{D}_k$. Then no edge $\{d, e\} \in E_{\text{cover}}$ exists for any $e \in E_k$ (Note that $E_k = \emptyset$ is possible). Hence (CSL(k)) is infeasible.
2. Now assume that (CSL) is feasible and let S^* be an optimal solution of (CSL). Define $S_k = S^* \cap E_k$. Then S_k is feasible for (CSL(k)):
   Consider $d \in \mathcal{D}_k$. Since S^* is feasible, we know that $d \in \text{cover}(S^*)$. In particular, there exists $s \in S^*$ such that $d \in \text{cover}(s)$, hence $s \in \mathcal{T}(d)$. Let $e = g(s) \in E$ be the edge containing s. This means, $e \cap \mathcal{T}(d) \neq \emptyset$ yielding that $\{d, e\} \in E_{\text{cover}}$. Due to point 3. in Notation 3.26 we have $e \in E_k$. From this we finally conclude $s \in e \in E_k$, hence $s \in S_k$, i.e., d is covered by some point in S_k.
   Now let S_k^* be an optimal solution of (CSL(k)). Since S_k is feasible for (CSL(k)) we obtain
   $$f_{\text{time}}(S_k^*) \leq f_{\text{time}}(S_k).$$
   On the other hand, $\cup_{k=1,\ldots,K} S_k^*$ is feasible for (CSL), since
   $$\text{cover}(\bigcup_{k=1,\ldots,K} S_k^*) = \bigcup_{k=1,\ldots,K} (\text{cover}(S_k^*)) = \bigcup_{k=1,\ldots,K} \mathcal{D}_k = \mathcal{D}.$$

Thus we obtain

$$\begin{aligned} f_{\text{time}}(\cup_{k=1,\dots,K} S_k^*) &\leq \sum_{k=1}^{K} f_{\text{time}}(S_k^*) \\ &\leq \sum_{k=1}^{K} f_{\text{time}}(S_k) \\ &= f_{\text{time}}(\bigcup_{k=1,\dots,K} S_k) \\ &= f_{\text{time}}(S^*) \quad \text{since the } S_k \text{ are pairwise disjoint.} \end{aligned}$$

This yields the optimality of $\bigcup_{k=1,\dots,K} S_k^*$. □

Decomposing (CSL) at least leads to smaller subproblems. In the case of the following corollary, we even get subproblems with a coefficient matrix with consecutive ones property.

Corollary 3.29. *Given $G = (V, E)$ with its set of points $\mathcal{T}$, as well as a set $\mathcal{D}$, suppose that each demand point in $\mathcal{D}$ can be covered by exactly one edge $e \subseteq \mathcal{T}$, i.e.,*

$$|\{e \in E : \gamma_d(d, e) \leq r\}| = 1 \text{ for all } d \in \mathcal{D}.$$

Then (CSL) can be decomposed into $|E|$ smaller problems (CSL(e)), $e \in E$, each of them containing only edge $e \in E$ as set of tracks. I.e., the input data of (CSL(e)) is $\mathcal{D}_e = \text{cover}_{\mathcal{D}}(e)$ and $\mathcal{T}_e = e$. Furthermore, the coefficient matrix of each subproblem has the consecutive ones property, and also the coefficient matrix of (CSL) has.

Proof. Consider $d \in \mathcal{D}$. Due to the assumptions of the corollary, there exists exactly one edge e with $\mathcal{T}(d) \cap e \neq \emptyset$, hence there exists exactly one edge incident with d in the cover graph G_{cover}. This means that no two edges e_1, e_2 are in the same connected component of G_{cover}, hence the problem decomposes into $|E|$ independent subproblems according to Theorem 3.28. Furthermore, since each subproblem involves only one single edge, we can apply Lemma 3.11, which implies that the consecutive ones property is satisfied for each coefficient matrix A_e^{cov} of subproblem (CSL(e)). Finally, the covering matrix A^{cov} of (CSL) can be written as

$$A^{\text{cov}} = \begin{pmatrix} A_{e_1}^{\text{cov}} & & & \\ & A_{e_2}^{\text{cov}} & & \\ & & \cdots & \\ & & & A_{e_{|E|}}^{\text{cov}} \end{pmatrix}$$

and hence also has the consecutive ones property. □

We remark that in practice the assumptions of Corollary 3.29 are often "almost" satisfied. The reason for this is that in practice the problem to be considered is not (SL) but originally (SL') (see page 19) referring to the establishment of additional stations, with a usually large set of already given existing stations S^{ex}. In the corresponding data set the existing stations usually coincide with the endpoints of the edges, i.e., $S^{ex} \subseteq V$ and $V \setminus S^{ex}$ is rather small. Now consider a node $v \in V$ with incident edges e_i and e_j. If there exists a station in v, a large portion of

$$\text{cover}(e_i) \cap \text{cover}(e_j)$$

is already covered. Hence, a situation as depicted in Figure 3.8 (see page 41) becomes less likely, since all three demand points would have been already covered by an existing station in v_0.

For an illustration of the practical applicability, see also the example depicted in Figure 3.11. In this example, the coefficient matrix of the complete problem has the consecutive ones property.

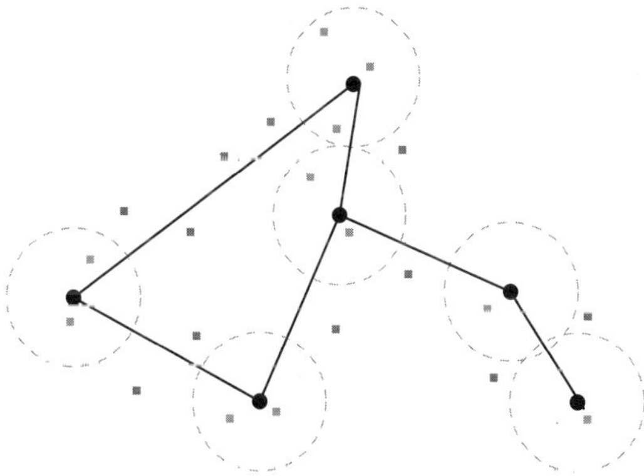

Fig. 3.11. An instance of (CSL) with consecutive ones property, since the lighter demand points are already covered by existing stations.

These theoretical considerations are confirmed by our numerical results (see [RS04, Ruf02]) which show that in the practical data set of *Deutsche Bahn* (see Section 2.1) the number of rows in which the ones do not appear consecutively is relatively small. For example, if we assume a covering radius of 2 km, there exist 1196 demand points which are not yet covered by an existing station, but which can be covered by some stop in $\mathcal{T}$, i.e.,

$$|(\mathcal{D} \setminus \text{cover}(S^{ex})) \cap \text{cover}(\mathcal{T})| = 1196.$$

This leads to 1196 rows in the matrix A^{cov}, and in only 299 of them the ones do not appear consecutively. This number can further be reduced to 148 by applying a permutation heuristic to obtain a better order of the columns (see

[Ruf02]). For a covering radius of only 1 kilometer, i.e., $r = 1$, the number of rows of A^{cov} only is 757. I.e., there exists 757 demand points which are closer to the tracks than 1 km, but the distance to their closest station is larger than 1 km. After applying the permutation heuristic to this example we are left with only seven rows not having the consecutive ones property. Note that in both examples, we further found out that the maximal number of blocks of consecutive ones was always less than or equal to 3, i.e., even if a row does not have the consecutive ones property, it can be decomposed into at most three parts in which the ones appear consecutively. This observation will be important in the next section, in which we show how problems with "almost" consecutive ones property can be solved.

3.6 Set Covering With Almost Consecutive Ones Property

As pointed out at the end of the previous section, in the practical data of the complete continuous stop location problem the number of rows of A^{cov} in which the ones do not appear consecutively is relatively small. This gives rise to developing a procedure for solving set covering problems in which the covering matrix "almost" has the consecutive ones property. Parts of this section have been obtained in [RS04], see also [Ruf02].

As in Section 3.4 the analysis presented here does not rely on any special property of the stop location problem but can be applied to any set covering problem. Consider

(SCP)

$$\begin{aligned} \min \quad & cx \\ \text{s.t. } A^{\text{cov}}x &\geq \mathbf{1}_{|\mathcal{D}|} \\ x &\in \{0,1\}^{|\mathcal{S}|}, \end{aligned} \tag{3.4}$$

and assume that A^{cov} "almost" has the consecutive ones property, i.e., in many rows of A^{cov} the 1s appear consecutively. Throughout the whole section we also assume that A^{cov} does **not** contain any zero row, i.e., that (SCP) is feasible. The goal is to find an optimal solution x^*, or equivalently, an optimal set $S^* \subseteq \mathcal{S}$ of columns of A^{cov}, where $S^* = \{s \in \mathcal{S} : x^*_s = 1\}$. The idea is to decompose the "bad" rows into a set of new rows, all of them satisfying that the ones appear consecutively, and to require that at least one of these rows needs to be covered. More precisely, we define:

Definition 3.30. *Let A^{cov} be a 0-1-matrix with $|\mathcal{D}|$ rows and $|\mathcal{S}|$ columns.*

1. *If A^{cov}_d is a row of A^{cov} let bl_d be its number of blocks of consecutive ones.*
2. *A^{cov}* **almost** *has the* **consecutive ones property** *, if $\sum_{d=1}^{|\mathcal{D}|} bl_d \ll |\mathcal{D}||\mathcal{S}|$.*

Now consider a matrix A^{cov} with $|\mathcal{D}|$ rows, such that in rows $1, \ldots, |\mathcal{D}| - p$ the 1s appear consecutively (i.e., $bl_d = 1$ for $d = 1, \ldots, |\mathcal{D}| - p$), and in rows $|\mathcal{D}| - p + 1, \ldots, |\mathcal{D}|$ we have $bl_d > 1$.

Notation 3.31. *Let A^{cov} be a 0-1-matrix and let bl_d be the number of blocks of consecutive ones in row d. For the ith block of consecutive ones in row d let*

- *$f_{d,i}$ be the column of the first 1 of block i and*
- *$l_{d,i}$ be the column of its last 1.*

This means, that

$$a_{dj} = \begin{cases} 1 & \text{if there exists } i \in \{1, \ldots, bl_d\} \text{ such that } f_{d,i} \leq j \leq l_{d,i} \\ 0 & \text{otherwise.} \end{cases}$$

Consider a row d with $bl_d > 1$. We replace A_d^{cov} by bl_d rows,

$$B_{d,1}, B_{d,2}, \ldots, B_{d,bl_d}$$

each of them containing only one single block, i.e., we define the jth element of row $B_{d,i}$ as

$$(B_{d,i})_j = \begin{cases} 1 & \text{if } f_{d,i} \leq j \leq l_{d,i} \\ 0 & \text{otherwise.} \end{cases}$$

The set covering problem

(SCP)

$$\begin{aligned} \min \quad & cx \\ s.t. \quad & A_d^{\text{cov}} x \geq 1 \quad \text{for } d = 1, \ldots, |\mathcal{D}| \\ & x \in \{0,1\}^{|\mathcal{S}|} \end{aligned}$$

can hence be reformulated as **(SCP')**

$$\begin{aligned} \min \quad & cx \\ s.t. \quad & A_d^{\text{cov}} x \geq 1 \quad \text{for } d = 1, \ldots, |\mathcal{D}| - p \\ & B_{d,i} x \geq y_{d,i} \quad \text{for } d = |\mathcal{D}| - p + 1, \ldots, |\mathcal{D}|, i = 1, \ldots, bl_d \\ & \textstyle\sum_{i=1}^{bl_d} y_{d,i} \geq 1 \quad \text{for } d = |\mathcal{D}| - p + 1, \ldots, |\mathcal{D}| \\ & y_{d,i} \in \{0,1\} \quad \text{for } d = |\mathcal{D}| - p + 1, \ldots, |\mathcal{D}|, i = 1, \ldots, bl_d \\ & x \in \{0,1\}^{|\mathcal{S}|}. \end{aligned}$$

Lemma 3.32. *(SCP) and (SCP') are equivalent.*

Proof.

(SCP) $\Longrightarrow$ (SCP'): Let x be a feasible solution of (SCP), i.e, $A^{\text{cov}} x \geq \mathbf{1}_{|\mathcal{D}|}$. We directly obtain that $Ax \geq \mathbf{1}_{|\mathcal{D}|-p}$. Moreover, for each row A_d^{cov}, $d = |\mathcal{D}| - p + 1, \ldots, |\mathcal{D}|$ we also know $A_d^{\text{cov}} x \geq 1$, i.e., there exists (at least) one block $i = \mathfrak{l}(d)$ of row d such that $B_{d,i} x \geq 1$. Defining

$$y_{d,i} = \begin{cases} 1 & \text{if } i = \mathfrak{l}(d) \\ 0 & \text{otherwise} \end{cases}$$

yields $B_{d,i}x \geq y_{d,i}$ and $\sum_{i=1}^{bl_d} y_{d,i} \geq y_{d,\mathfrak{l}(d)} = 1$, hence (x, y) is feasible for (SCP') with the same objective value.

(SCP') $\Longrightarrow$ (SCP): On the other hand, each feasible solution of (SCP') satisfies $A_d^{\text{cov}}x \geq 1$ for $d = 1, \ldots, |\mathcal{D}| - p$, while for $d = |\mathcal{D}| - p + 1, \ldots, |\mathcal{D}|$ we know that

$$\sum_{i=1}^{bl_d} y_{d,i} \geq 1$$

and hence there exists (at least) one $i = \mathfrak{l}(d)$ for each row d with $y_{d,\mathfrak{l}(d)} = 1$. From this we conclude

$$B_{d,\mathfrak{l}(d)}x \geq y_{d,\mathfrak{l}(d)} = 1,$$

i.e., x covers block $l = \mathfrak{l}(d)$ of row d. This finally yields $A_d^{\text{cov}}x \geq 1$ also for $d = |\mathcal{D}| - p + 1, \ldots, |\mathcal{D}|$. Together, $A^{\text{cov}}x \geq 1$, hence x is feasible for (SCP) with the same objective value. □

It is more convenient to rewrite (SCP') in matrix form. To this end, we define

- the matrix A as the first $|\mathcal{D}| - p$ rows of A^{cov},
- $bl = \sum_{d=|\mathcal{D}|-p+1}^{|\mathcal{D}|} bl_d$ as the total number of blocks in rows of A^{cov} without consecutive ones,
- B as the matrix containing the bl rows $B_{d,i}$, and
- C as a matrix with p rows and bl columns, such that row i of C is defined by

$$C_{ij} = \begin{cases} 1 & \text{if } \sum_{d=|\mathcal{D}|-p+1}^{|\mathcal{D}|-p+i-1} bl_d < j \leq \sum_{d=|\mathcal{D}|-p+1}^{|\mathcal{D}|-p+i} bl_d \\ 0 & \text{otherwise.} \end{cases}$$

In the following we will use the next – equivalent – formulation of (SCP'):

(SCP')

$$\begin{aligned} \min\ & cx \\ \text{s.t. } & Ax \geq \mathbf{1}_{|\mathcal{D}|-p} \\ & Bx - Iy \geq \mathbf{0}_{bl} \\ & Cy \geq \mathbf{1}_p \\ & x \in \{0,1\}^{|\mathcal{S}|}, \\ & y \in \{0,1\}^{bl}. \end{aligned}$$

The constraint $Cy \geq \mathbf{1}_p$ makes sure that at least one block of each row A_d^{cov} with $d \geq |\mathcal{D}| - p + 1$ is covered.

There are several advantages of this reformulation. The first is that all three matrices A,B, and C have the consecutive ones property. But note that the coefficient matrix of (SCP') does not have the consecutive ones property, and also is not totally unimodular, such that in general non-integer basic solutions exist. How the new formulation can be used to find good bounds and to set up a branch and bound approach is described next. It can also be used to provide an approximation algorithm with an approximation ratio of $\max_{d=1,\dots,|\mathcal{D}|} bl_d$ as done in [MSW05].

Lower Bounds

A lower bound on (SCP') is obtained by relaxing all constraints that contain variables $y_{d,i}$, yielding a problem with a coefficient matrix with consecutive ones property. For the stop location problem this can be interpreted as simply forgetting about the demand points which destroy the consecutive ones property of the matrix, i.e., we do not require to cover them. The corresponding IP is

(SCPl)

$$\begin{aligned} \min\ & cx \\ \text{s.t. } & Ax \geq \mathbf{1}_{|\mathcal{D}|-p} \\ & x \in \{0,1\}^{|\mathcal{S}|}. \end{aligned}$$

Lemma 3.33. *Each optimal solution of (SCPl) is a lower bound on (SCP).*

Proof. Since A only contains a part of the rows of A^{cov}, (SCPl) is a relaxation of (SCP), and the result follows. □

Since A has the consecutive ones property, a solution of (SCPl) can be calculated efficiently by one of the approaches discussed in Section 3.4. A better lower bound can be found by considering the dual of the LP-relaxation of (SCP'). Since there exists an optimal solution with $x_s \leq 1$ for $s \in \mathcal{S}$ and $y_d \leq 1$ for $d \in \mathcal{D}$, the dual of the LP-relaxation is given by

(Dual-SCP')

$$\begin{aligned} \max\ & \mathbf{1}_{|\mathcal{D}|-p}\eta_A + \mathbf{1}_{bl}\eta_C \\ \text{s.t. } & A^T\eta_A + B^T\eta_B \leq c && (3.5)\\ & -\eta_B + C^T\eta_C \leq 0 && (3.6)\\ & \eta_A, \eta_B, \eta_C \geq 0. && (3.7) \end{aligned}$$

A feasible solution of (Dual-SCP') is obtained by solving the following sequence of problems.

1. Solve problem **(A)**

$$\begin{aligned} \max\ & \mathbf{1}\eta_A \\ \text{s.t. } & A^T\eta_A \leq c \\ & \eta_A \geq 0. \end{aligned}$$

   Let η_A^* be an optimal solution of (A).

2. Solve problem **(B)**

$$\begin{aligned} &\max \mathbf{1}\eta_B \\ &\text{s.t. } B^T\eta_B \le c - A^T\eta_A^* \\ &\eta_B \ge 0. \end{aligned}$$

Let η_B^* be an optimal solution of (B).

3. Solve problem **(C)**

$$\begin{aligned} &\max \mathbf{1}\eta_C \\ &\text{s.t. } C^T\eta_C \le \eta_B^* \\ &\eta_C \ge 0. \end{aligned}$$

Let η_C^* be an optimal solution of (C).

The following properties apply.

Lemma 3.34.

1. *(A), (B), and (C) are feasible.*
2. *$(\eta_A^*, \eta_B^*, \eta_C^*)$ is a feasible solution of (Dual-SCP').*
3. *$\mathbf{1}_{|\mathcal{D}|-p}\eta_A + \mathbf{1}_{bl}\eta_C$ is a lower bound on (SCP').*

Proof.

1. (A) is feasible since $c \ge 0$ and hence $\eta_A = 0$ solves the problem. Since η_A^* is feasible for (A) it satisfies

$$c - A^T\eta_A^* \ge 0$$

and hence $\eta_B = 0$ is a feasible solution of (B) for any feasible solution η_A^* of (A). Similarly, since $\eta_B^* \ge 0$ a feasible solution of (C) is given by $\eta_C^* = 0$.

2. Constraint (3.5) is satisfied since from the feasibility of η_B^* for (B) we know that

$$A^T\eta_A^* + B^T\eta_B^* \le c.$$

Furthermore, η_C^* is feasible for (C), i.e.,

$$-\eta_B^* + C^T\eta_C \le 0,$$

which is constraint (3.6) of (Dual-SCP').

3. This follows from the feasibility of $(\eta_A^*, \eta_B^*, \eta_C^*)$ for (Dual-SCP'), see part 2. □

The next lemma shows that this second bound is always better than the first bound we obtained by solving (SCPl).

Lemma 3.35. *Let x^l be the optimal solution of (SCPl). Then $cx^l \le \mathbf{1}_{|\mathcal{D}|-p}\eta_A^* + \mathbf{1}_{bl}\eta_C^*$, where $\eta_A^*, \eta_B^*, \eta_C^*$ are optimal solutions for (A), (B), (C), respectively.*

Proof. Note that the dual of (A) is (SCPl). Hence, from the strong duality theorem, see, e.g., [HK01, NW88] we know that

$$cx^l = \mathbf{1}_{|\mathcal{D}|-p}\eta_A^*,$$

hence $cx^l \leq \mathbf{1}_{|\mathcal{D}|-p}\eta_A^* + \mathbf{1}_{bl}\eta_C^*$. □

Moreover, this bound can also be calculated efficiently by using that A^T, B^T, and C^T are interval matrices and solving the corresponding problems (A),(B), and (C) along the lines of Section 3.4.

Upper Bounds

Fixing all $y_{d,i} = 1$ in (SCP') also results in a problem in which the coefficient matrix has the consecutive ones property. Moreover, it yields a feasible solution to the original problem. In terms of the stop location problem this strategy requires that each demand point d which can be covered by more than one edge *must* be covered by stops on all possible edges. The solution found is hence feasible but will in general have too many new stations opened. Formally, this solution is found by solving

(SCPu1)

$$\begin{aligned} \min\ & cx \\ \text{s.t. } & Ax \geq \mathbf{1}_{|\mathcal{D}|-p} \\ & Bx \geq \mathbf{1}_{bl} \\ & x \in \{0,1\}^{|\mathcal{S}|}. \end{aligned}$$

Lemma 3.36. *Each feasible solution of (SCPu1) is an upper bound on (SCP').*

Proof. Let x^u be a feasible solution of (SCPu1), and x^*, y^* be optimal for (SCP'). Defining $y = \mathbf{1}_{bl}$ yields a feasible solution (x^u, y) of (SCP'), hence satisfying $cx^u \geq cx^*$. □

A better upper bound is obtained if we do not require that all rows of B are covered, but select only one of them for each demand point d.

Notation 3.37. *Let $\mathfrak{l}(d) : \{|\mathcal{D}| - p + 1, \ldots, |\mathcal{D}|\} \to \mathbb{N}$ be a mapping selecting a block $i = \mathfrak{l}(d)$ for each row $d \in \{|\mathcal{D}| - p + 1, \ldots, |\mathcal{D}|\}$. We call the mapping $\mathfrak{l}$ feasible if*

$$1 \leq \mathfrak{l}(d) \leq bl_d$$

for all $d = |\mathcal{D}| - p + 1, \ldots, |\mathcal{D}|$. We also write $\mathfrak{l} \subseteq \{|\mathcal{D}| - p + 1, \ldots, |\mathcal{D}|\} \times \mathbb{N}$ to specify $\mathfrak{l}$.

We obtain the following result: Each feasible solution of the following program

(SCPu($\mathfrak{l}$))

$$\begin{array}{rl} \min & cx \\ \text{s.t.} & Ax \geq \mathbf{1}_{|\mathcal{D}|-p} \\ & B_{d,\mathfrak{l}(d)}x \geq 1 \quad \text{for all } |\mathcal{D}|-p+1,\ldots,|\mathcal{D}| \\ & x \in \{0,1\}^S \end{array}$$

gives an upper bound on (SCP'). Moreover, the best bound obtained by solving (SCPu($\mathfrak{l}$)) is better than the best bound obtained by solving (SCPu1).

Lemma 3.38. *Let x^* be the optimal solution of (SCP).*

1. *Each feasible solution x of (SCPu($\mathfrak{l}$)) satisfies $cx \geq cx^*$.*
2. *If x^{u1} is an optimal solution of (SCPu1), and x^{u2} an optimal solution of (SCPu($\mathfrak{l}$)) (for any feasible mapping $\mathfrak{l}$) we have*

$$cx^* \leq cx^{u2} \leq cx^{u1}.$$

Proof.

1. We define for $d = |\mathcal{D}| - p + 1, \ldots, |\mathcal{D}|$

$$y_{d,i} = \begin{cases} 1 & \text{if } i = \mathfrak{l}(d) \\ 0 & \text{otherwise} \end{cases}$$

   to obtain a feasible solution (x, y) for (SCP') with the same objective value as (SCPu($\mathfrak{l}$)).
2. $cx^* \leq cx^{u2}$ directly follows from part 1 of this lemma, while $cx^{u2} \leq cx^{u1}$ holds since (SCPu($\mathfrak{l}$)) is a relaxation of (SCPu1). □

Heuristic Approaches

In the following we suggest two heuristics for (SCP'). Both work by choosing a good mapping $\mathfrak{l}(d)$ for the formulation (SCP($\mathfrak{l}$)).

The first one is based on a cost-argument, i.e., for each row we choose the cheapest block that can be used to cover the row. The interpretation for the stop location problem is the following: We require to cover each demand point from that edge $e \in E$ of the graph (V, E) with the lowest traffic load c_e.

The first heuristic works as follows.

Heuristic 6: Cost-Heuristic for (SCP)

Input: A^{cov}, b, c.
Output: A feasible solution x of (SCP).
Step 1. Obtain matrices A and B of (SCP').
Step 2. For $d = |\mathcal{D}| - p + 1, \ldots, |\mathcal{D}|$:
  Assign $\mathfrak{l}(d) = i$ if $c_j = \min_{j': a_{dj'}=1} c_{j'}$ and $f_{d,i} \leq j \leq l_{d,i}$.

```
Step 3. Let x,y be the solution of (SCPu(l)) (e.g., by Algorithm 5).
Step 4. Output: x.
```

In our second heuristic we do not focus on the costs c but choose that block for row d which can be used to cover most (other) demand points, i.e., we choose the block containing the candidate with the largest coverage in the row. The formal description is the following.

Heuristic 7: Coverage-Heuristic for (SCP)

Input: A^{cov}, b, c.
Output: A feasible solution x of (SCP).
Step 1. Obtain matrices A and B of (SCP').
Step 2. For $d = |\mathcal{D}|-p+1, \ldots, |\mathcal{D}|$: Assign $\mathfrak{l}(d) = l$ if $\max_{j':a_{dj'}=1} |\text{cover}(j')| = |\text{cover}(j)|$ and $f_{d,i} \leq j \leq l_{d,i}$.
Step 3. Let x, y be the solution of (SCPu($\mathfrak{l}$)), (e.g., by Algorithm 5).
Step 4. Output: x.

Apart from these two simple heuristics we can also construct a feasible solution and an upper bound by combining the lower bound obtained from (SCPl) (or equivalently, by solving the dual program (A)) with the cost-based heuristic (Algorithm 6). To this end, let x^l be an optimal solution of (SCPl). Then determine the set of rows which are not covered by x^l, i.e., define $\mathfrak{D} = \{d : A_d^{\text{cov}} x^l = 0\}$ and choose $\mathfrak{l}(d)$ according to Algorithm 6 for all $d \in \mathfrak{D}$. Then solve the reduced set covering problem

(Red-SCP($\mathfrak{l}$))

$$\begin{aligned} \min \quad & cx \\ \text{s.t. } & B_{d,\mathfrak{l}(d)} x \geq 1 \text{ for all } d \in \mathfrak{D} \\ & x \in \{0,1\}^{|\mathcal{S}|} \end{aligned}$$

and let $\tilde{x}$ be an optimal solution. By defining

$$x_s = \max\{x_s^l, \tilde{x}_s\}$$

we then obtain an upper bound on (SCP').

Lemma 3.39. *Let x^* be an optimal solution of (SCP). Furthermore, let x^l be an optimal solution of (SCPl) and $\tilde{x}$ be an optimal solution of (Red-SCP($\mathfrak{l}$)). Then x defined by*

$$x_s = \max\{x_s^l, \tilde{x}_s\}, \; s = 1, \ldots, |\mathcal{S}|$$

satisfies $cx \geq cx^*$.

Proof. We only have to show that x is feasible for (SCP). Let

$$\mathfrak{D} = \{d : A_d^{\text{cov}} x^l = 0\}.$$

Then, for all $d \notin \mathfrak{D}$ we have that

$$A_d^{\text{cov}} x \geq A_d^{\text{cov}} x^l \geq 1,$$

hence these rows are covered by x. Now take $d \in \mathfrak{D}$. Then

$$A_d^{\text{cov}} x \geq A_d^{\text{cov}} \tilde{x} \geq B_{d,\mathfrak{l}(d)} \tilde{x} \geq 1.$$

Together, $A^{\text{cov}} x \geq \mathbf{1}_{|\mathcal{D}|}$ and the result follows. □

The above discussion leads to the following algorithm that contains upper and lower bound computation.

Algorithm 8: Upper and lower bound for (SCP)

Input: A^{cov}, b, c.
Output: Upper bound cx^u and lower bound f^l on (SCP).
Step 1: Derive the matrices A, B, C of (SCP') and solve (A),(B), and (C) with optimal solutions $\eta_A^*, \eta_B^*, \eta_C^*$.
 Let x^A be the dual solution of (SCP1).
Step 2: Define $\mathfrak{D} = \{d : A_d^{\text{cov}} x^A = 0\}$.
Step 3: Calculate $\mathfrak{l}(d)$ for all $d \in \mathfrak{D}$ according to Algorithm 6.
Step 4: Solve (Red-SCP($\mathfrak{l}$)) with respect to $\mathfrak{D}$ and $\mathfrak{l}$.
 Let $\tilde{x}$ be the solution.
Step 5: Define for all $s = 1, \ldots, |S|$: $x_s^u = \max\{x_s^A, \tilde{x}_s\}$.
Step 6: Output: x^u and $f^l = \mathbf{1}_{|\mathcal{D}|-p}\eta_A + \mathbf{1}_{bl}\eta_C$.

Branch and Bound Approach

For solving (SCP') we propose a branch and bound algorithm. The idea is to consider a row d (for $|\mathcal{D}| - p < d \leq |\mathcal{D}|$) in each layer of the branch and bound tree and iteratively select one of the y_{di} and set it to one. This means, the corresponding row B_{di} can be added to matrix A while all other rows $B_{di'}$ with $i' \neq i$ can be deleted from B. Formally, we obtain:

Notation 3.40. *Let $\mathfrak{D} \subseteq \{|\mathcal{D}| - p + 1, \ldots, |\mathcal{D}|\}$ be a set of rows with $bl_d > 1$ for all $d \in \mathfrak{D}$. Moreover, let $\mathfrak{l}$ be a feasible mapping, i.e., $\mathfrak{l}(d) \in \{1, \ldots, bl_d\}$ for $d \in \mathfrak{D}$. For a given instance of (SCP') define $P(\mathfrak{D}, \mathfrak{l})$ as an instance of (SCP') in which $y_{d,\mathfrak{l}(d)} = 1$ for all $d \in \mathfrak{D}$.*

Using the notation $\mathfrak{D}^C = \{|\mathcal{D}| - p + 1, \dots, |\mathcal{D}|\} \setminus \mathfrak{D}$ we get

P($\mathfrak{D}, \mathfrak{l}$)

$$\begin{array}{lr}
\min & cx \\
\text{s.t.} & Ax \geq \mathbf{1}_{|\mathcal{D}|-p} \\
& B_{d,\mathfrak{l}(d)}x \geq 1 \text{ for all } d \in \mathfrak{D} \\
& B_{d,i}x \geq y_{d,i} \ \text{ for } d \in \mathfrak{D}^C, i = 1, \dots, bl_d \\
& \sum_{i=1}^{bl_d} y_{d,i} \geq 1 \ \text{ for } d \in \mathfrak{D}^C \\
& y_{d,i} \in \{0,1\} \ \text{ for } d \in \mathfrak{D}^C, i = 1, \dots, bl_d \\
& x \in \{0,1\}^{|\mathcal{S}|}.
\end{array}$$

Lemma 3.41. *Let x^* be an optimal solution of (SCP), and let $x^{\mathfrak{D},\mathfrak{l}}, y^{\mathfrak{D},\mathfrak{l}}$ be an optimal solution of P($\mathfrak{D}, \mathfrak{l}$). Then*

1. $cx^* \leq cx^{\mathfrak{D},\mathfrak{l}}$.
2. *For each fixed $\mathfrak{D} \subseteq \{|\mathcal{D}| - p + 1, \dots, |\mathcal{D}|\}$ we have*

$$cx^* = \min_{\mathfrak{l} \text{ feasible}} cx^{\mathfrak{D},\mathfrak{l}}.$$

Proof.

1. Taking the solution $x^{\mathfrak{D},\mathfrak{l}}, y^{\mathfrak{D},\mathfrak{l}}$ and setting for all $d \in \mathfrak{D}$

$$y_{d,i} = \begin{cases} 1 & \text{if } i = \mathfrak{l}(d) \\ 0 & \text{otherwise} \end{cases}$$

yields a feasible solution of (SCP') and hence $cx^* \leq cx^{\mathfrak{D},\mathfrak{l}}$.

2. Let x^*, y^* be an optimal solution of (SCP'). Then for all $d \in \{|\mathcal{D}| - p + 1, \dots, |\mathcal{D}|\}$ there exists some i such that $y_{di} = 1$. Define $\mathfrak{l}(d) = i$ for all $d \in \mathfrak{D}$ and let $y^{\mathfrak{D}}$ be the vector y^*, restricted to the components of $\mathfrak{D}$. This means, $x^*, y^{\mathfrak{D}}$ is feasible for P($\mathfrak{D}, \mathfrak{l}$) and consequently,

$$cx^{\mathfrak{D},\mathfrak{l}} \leq cx^*$$

From part 1 we already know $cx^* \leq cx^{\mathfrak{D},\mathfrak{l}}$, hence equality is attained. □

The following observations are the basis for the branch and bound approach.

- $P(\emptyset, \emptyset) =$(SCP').
- Fixing $y_{di} = 1$ in $P(\mathfrak{D}, \mathfrak{l})$ for some $d \in \mathfrak{D}^C$ and for some $1 \leq i \leq bl_d$ leads to $P(\mathfrak{D} \cup \{d\}, \mathfrak{l} \cup \{d, i\})$.
- The coefficient matrix of $P(\{|\mathcal{D}| - p + 1, \dots, |\mathcal{D}|\}, \mathfrak{l})$ has the consecutive ones property and the problem can hence be solved efficiently by Algorithm 5.

Thus, by iteratively fixing variables y_{di} we always obtain subproblems of the same type, and in each iteration the number of rows d with $bl_d = 1$ increases (yielding a larger matrix A with consecutive ones property) while the number of "bad" rows d with $bl_d > 1$ decreases. Hence, we get closer to the consecutive ones property in each step. Before we formulate the branch and bound procedure we remark that we can reduce the size of (SCP') by applying the following reduction rules.

Lemma 3.42.

1. *If* $T(d_1) \subseteq T(d_2)$ *for* $1 \leq d_1 \leq |\mathcal{D}| - p$ *then a solution of (SCP') can be found by considering the reduced problem without rows* $B_{d_2,i}$ *for all* $i = 1, \ldots, bl_{d_2}$.
2. *If* $T(d_1) = \{s\}$ *for some* $1 \leq d_1 \leq |\mathcal{D}| - p$, *then in all optimal solutions,* $x_s = 1$, *and it is sufficient to consider the reduced problem without column* s *and without all rows* $B_{d_2,i}$ *with* $d_2 \in \text{cover}(s)$ *and* $1 \leq i \leq bl_{d_2}$.

Proof. The rules are transfered from Lemma 3.16 taking into account that not all rows of the matrix B need to be covered, i.e., we require that rows denoted by d_1 in the formulation of Lemma 3.16 are rows of the matrix A. □

The branch and bound algorithm can finally be stated as follows.

Algorithm 9: Branch and bound for (SCP)

Input: A^{cov}, b, c, and accuracy ϵ.
Output: Feasible solution x of (SCP). such that $|cx - cx^*| \leq \epsilon$, if x^* is the optimal objective value.
Step 0. Set $\mathfrak{D} = \emptyset$, $\mathfrak{l} = \emptyset$, and $S^* = \mathcal{S}$.
    Derive P by reducing $P(\mathfrak{D}, \mathfrak{l})$ according to Lemma 3.42.

$$f^{\mathbf{l}}, f^{\mathbf{u}} = \text{ lower and upper bound, obtained by Algorithm 8,}$$
$$\text{List} = \{P\} \text{ with lower bound } f_P^l = f^{\mathbf{l}}.$$

Step 1.
    1. If List$= \emptyset$, stop: Exact optimal solution is x^*.
    2. $f^{\mathbf{l}} = \min\{f_P^l : P \in \text{List}\}$
    3. If $f^{\mathbf{u}} - f^{\mathbf{l}} \leq \epsilon$ stop: ϵ optimal solution is x^*.
Step 2. Choose $P = P(\mathfrak{D}, \mathfrak{l}) \in \text{List}$ with current lower bound f_P^l.
Step 3. **Reduction of P:** Check if P can be reduced according to Lemma 3.42.
    (Only the latest added row $d \in \mathfrak{D}$ needs to be considered as d_1)
Step 4. **Bounds** Use Algorithm 8 to obtain a new lower bound f_P^l and a feasible solution x^u.
Step 5. **Pruning**
    1. If $f_P^l = cx_P^u$, prune by optimality, i.e.,

$$x^* = x_P^u \text{ if } cx_P^u < f^{\mathbf{u}}$$
$$f^{\mathbf{u}} = \min\{cx_P^u, f^{\mathbf{u}}\}$$
$$\text{List} = \text{List} \setminus \{P\}.$$

    Goto 1.
    2. If $f_P^l \geq f^{\mathbf{u}}$ prune by bound, i.e., $\text{List} = \text{List} \setminus \{P\}$.
    Goto 1.
    3. If $cx_P^u < f^{\mathbf{u}}$ set

$$f^{\mathbf{u}} = cx_P^u$$
$$x^* = x_P^u.$$

`Step 6. Choose` $d \in \mathfrak{D}^C$ `and set`

$$\text{List} = \text{List} \cup \{P(\mathfrak{D} \cup \{d\}, \mathfrak{l} \cup (d,i)) : i \in \{1, \ldots, bl_d\}\}$$

`Goto 1.`

The application of this branch and bound approach on random data yields optimal solutions for 100×100 matrices A^{cov} in less than a minute on a standard personal computer, if the number of blocks of consecutive ones is not too high. As expected, the running time increases drastically with the number of blocks. A detailed analysis is presented in the diploma thesis of [Ruf02], see also [RS04].

We also applied Algorithm 9 to the real-world data described in Section 2.1. For a covering radius of $r = 2$, i.e., if we require that all demand points should be closer than 2 km to their nearest station, we obtained a solution with an optimality gap

$$\frac{f_{\text{time}}(x^u) - f_{\text{time}}(x^l)}{f_{\text{time}}(x^l)} \leq 0.017,$$

i.e., of less than 1.7 % in the first iteration of step 4 in the branch and bound algorithm. This shows that Algorithm 8 behaves very well in this case. The improvement of the starting solution by Algorithm 9, however, was rather slow. For the same problem instance, but with a covering radius of only 1 km, the initial optimality gap was only 0.1 %, such that in this case we can consider the problem as solved by Algorithm 8 only.

4

Bicriteria Stop Location

There are many useful extensions of the complete continuous stop location problem (CSL). In this chapter we investigate the following bicriteria variant. In a practical setting, one might not want to cover all demand points $\mathcal{D}$ but only a given percentage of the population. To this end we assume that for each demand point, we have given a weight w_d representing the number of customers who would like to use public transportation, if the next station was closer than r. For a given set of stops S, recall from Section 2.3 that $f_{\text{cover}}(S)$ denotes the number of (potential) customers who live closer than r to some stop in S. Certainly, it is preferable to cover as many customers as possible, i.e, to maximize $f_{\text{cover}}(S)$. On the other hand, establishing many new stops for trains is costly and increases the travel time for customers, because each stop needs an additional time of, e.g., two minutes.Hence, we also have to take care of the additional travel time f_{time} which can be computed as the total time used for the additional stopping activities summed over all customers. We are hence dealing with the following bicriteria variant of (SL).

(BSL)

Given $G = (V, E)$ with its set of points on edges $\mathcal{T} = \bigcup_{e \in E} e$ and with traffic loads c_e for all $e \in E$, and c_v for all $v \in V$, as well as a finite set of points $\mathcal{D}$ with weights w_d and gauges γ_d for all $d \in \mathcal{D}$, find a set $S \subseteq \mathcal{T}$ such that both

$$f_{\text{time}}(S) = \sum_{s \in S} c_{g(s)}, \quad \textit{and}$$

$$-f_{\text{cover}}(S) = -\sum_{d \in \text{cover}(S)} w_d$$

are minimized.

We remark that for a subset $S \subseteq \mathcal{T}$ the function $f_{\text{cover}}(S)$ was already used in Section 2.3. Also note that parts of Sections 4.1, 4.2, and of Section 4.3 have recently been published in [Sch05c].

Chapter 4 is structured as follows: We first present the bicriteria problem and its two e-constraint versions and point out their relation to efficient solutions. We then prove that the finite candidate set developed for (CSL) can also be used for the bicriteria problem. Since this leads to bicriteria set covering problems, we analyze such problems for the special case that the covering matrix has the consecutive ones property and prove their equivalence to bicriteria shortest path problems. Finally we present an alternative finite dominating set which can be used independent of the given radius r.

4.1 Constraint Problems and Lexicographic Minimality

What we mean by "minimizing both" objective functions is to find Pareto solutions of the problem with respect to f_{time} and f_{cover}. Recall from Appendix B that if $S_1, S_2 \subseteq \mathcal{T}$ denote two feasible sets of stops, S_1 dominates S_2 if

$$f_{\text{time}}(S_1) \leq f_{\text{time}}(S_2) \text{ and}$$
$$f_{\text{cover}}(S_1) \geq f_{\text{cover}}(S_2),$$

where at least one of these inequalities is strict. Then a *Pareto solution* S^* is a feasible set of stops which is not dominated by any other feasible set of stops (see Appendix B). The points $\begin{pmatrix} f_{\text{time}}(S^*) \\ f_{\text{cover}}(S^*) \end{pmatrix}$ for Pareto solutions S^* in the objective space are called *efficient* points.

To find Pareto solutions a common idea is to minimize only one of the two objective functions and to bound the other objective in the constraint set. Since this can be done for both objective functions, we obtain the following two one-criteria problems, which are called e-constraint problems in the literature.

(BSL-time) Given $\mathcal{D}$, $G = (V, E)$ with its set of points $\mathcal{T}$, weights c_e, c_v, w_d, gauges γ_d, and a lower bound $Q_{\text{cover}} \in \mathbb{R}$ on f_{cover}, find a set $S^* \subseteq \mathcal{T}$ such that $f_{\text{cover}}(S^*) \geq Q_{\text{cover}}$ and $f_{\text{time}}(S^*)$ is minimal.

(BSL-cover) Given $\mathcal{D}$, $G = (V, E)$ with its set of points $\mathcal{T}$, weights c_e, c_v, w_d, gauges γ_d, and an upper bound $Q_{\text{time}} \in \mathbb{R}$ on f_{time}, find a set $S^* \subseteq \mathcal{T}$ such that $f_{\text{time}}(S^*) \leq Q_{\text{time}}$ and $f_{\text{cover}}(S^*)$ is maximal.

Due to Haimes and Chankong [HC83] (see Appendix B) we have the following result, providing the connection between Pareto solutions and the optimal solutions of the e-constraint problems.

Lemma 4.1.

1. *Let S be a unique optimal solution of (BSL-*time*). Then S is a Pareto solution. If more than one optimal solution of (BSL-*time*) exists, the solutions that additionally maximize f_{cover} are Pareto solutions.*
2. *Let S be a unique optimal solution of (BSL-*cover*). Then S is a Pareto solution. If more than one optimal solution of (BSL-*cover*) exists, the solutions that additionally minimize f_{time} are Pareto solutions.*

Using Lemma 4.1 to find Pareto solutions is known as the *e-constraint method*, see, e.g., [Ehr00]. Unfortunately, both e-constraint problems are hard to solve.

Corollary 4.2. *(BSL) and the two e-constraint problems (BSL-*time*) and (BSL-*cover*) are NP-hard, even if all weights c_e, c_v, w_d are equal to 1.*

Proof. We already know that (CSL) is NP-hard under the conditions of the corollary, see Theorem 3.2. The decision version of both e-constraint problems (BSL-time) and (BSL-cover) is the following:

Given $\mathcal{D}$, $G = (V, E)$ with its set of points $\mathcal{T}$, weights c_e, c_v, w_d, gauges γ_d, and $Q_{\text{time}}, Q_{\text{cover}} \in \mathbb{R}$, does there exist a set $S^* \subseteq \mathcal{T}$ such that $f_{\text{time}}(S^*) \leq Q_{\text{time}}$ and $f_{\text{cover}}(S^*) \geq Q_{\text{cover}}$?

Defining $Q_{\text{cover}} = \sum_{d \in \mathcal{D}} w_d$ shows that the decision version of (CSL) is a special case of the decision version of both (BSL-time) and (BSL-cover) and thus both e-constraint problems are NP-hard. □

We now discuss the two lexicographic optimal solutions, for which we know that they are Pareto solutions (see Appendix B).

- Maximizing f_{cover} as first objective means that we have to cover **all** demand points, that can be covered, i.e., all demand points d with $\mathcal{T}(d) \neq \emptyset$. This yields exactly (CSL) of the previous chapter, if we define

$$\mathcal{D}' = \mathcal{D} \cap \text{cover}(\mathcal{T})$$

  as the set of demand points to cover. In Chapter 3 we have shown that this problem is NP-hard.
- On the other hand, minimizing f_{time} leads to a trivial problem since it can be solved easily by not installing any stop at all. To find a lexicographic minimal solution $S \subseteq \mathcal{T}$ we first determine the set of edges and nodes with zero costs, i.e.,

$$E^0 = \{e \in E : c_e = 0\}$$
$$V^0 = \{v \in V : c_v = 0\}.$$

  If both sets are empty, $S = \emptyset$ is the unique lexicographic minimal solution. In the case that we can locate stops with zero costs, we further determine

$$\mathcal{D}' = \text{cover}(E^0) \cup \text{cover}(V^0)$$

as the set of all demand points that can be covered with zero costs. Finally, we cover all demand points in $\mathcal{D}'$ as follows: For all $d \in \text{cover}(E^0)$ we select $e \in E^0$ such that $d \in \text{cover}(e)$ and choose

$$s(d) \in \text{argmin}_{s \in e} \gamma_d(d, s).$$

Then $V^0 \cup \{s(d) : d \in \mathcal{D}\}$ covers $\mathcal{D}'$ with zero costs and hence is a lexicographic minimal solution. In other words, we install a stop in each node of V^0 and in all the projection points of demand points on a 0-cost edge.

We mention that the unweighted version of (BSL-cover), i.e., to locate at most $K = Q_{\text{time}}$ stops in such a way that f_{cover} is maximized, was investigated in [KPS+03] for the case of one single straight-line track and for the case of two parallel straight-line tracks. For both cases, polynomial time algorithms using dynamic programming were developed with a time complexity of $O(K|\mathcal{D}|^2)$ for the single track case. Moreover, it is shown that along one straight line track, the unweighted version (BSL-cover) is equivalent to a one-dimensional uncapacitated and unimodular K-facility location problem. As observed by [Tam02] the problem can hence be solved in $O(K|\mathcal{D}|log(\mathcal{D}))$ time.

4.2 Integer Programming Formulations

We now use the methodology developed in Chapter 3 and again derive a finite dominating set. Recall the definition of the finite dominating set $\mathcal{S}$ for (CSL) given in Notation 3.5 and in (3.2) on page 27 in Section 3.2, i.e.,

$$\mathcal{S}^e = \{v_1^e, v_2^e\} \cup \begin{cases} \bigcup_{d \in \mathcal{D}} \{f_d^e, l_d^e\} & \text{if } \bigcup_{d \in \mathcal{D}} \{f_d^e, l_d^e\} \text{ contains at least one point of the interior of } e \\ \{\frac{v_1^e + v_2^e}{2}\} & \text{otherwise.} \end{cases}$$

where v_1^e, v_2^e are the endpoints of edge e and f_d^e, l_d^e are the endpoints of the interval $\mathcal{T}(d) \cap e$. Fortunately, the following theorem shows that

$$\mathcal{S} = \bigcup_{e \in E} \mathcal{S}^e$$

is also a finite dominating set for the bicriteria stop location problem.

Theorem 4.3. *$\mathcal{S}$ is a finite dominating set for (BSL-time), (BSL-cover), and for (BSL). More precisely,*

- *Either (BSL-time) is infeasible, or there exists an optimal solution $S^* \subseteq \mathcal{S}$.*
- *Either (BSL-cover) is infeasible, or there exists an optimal solution $S^* \subseteq \mathcal{S}$.*
- *Let $(Q_{\text{time}}, Q_{\text{cover}})$ be an efficient solution of (BSL). Then there exists a Pareto solution $S \in \mathcal{S}$ with $f_{\text{time}}(S) = Q_{\text{time}}$ and $f_{\text{cover}}(S) = Q_{\text{cover}}$.*

Proof. Given some optimal (or Pareto) set S^*, we use exactly the proof of Theorem 3.7 and construct a set $S' \subseteq \mathcal{S}$ by moving stops of the given set S^* into points of $\mathcal{S}$ without changing the objective function values. For S' we hence obtain

$$f_{\text{cover}}(S^*) \le f_{\text{cover}}(S') \quad \text{and}$$
$$f_{\text{time}}(S^*) \ge f_{\text{time}}(S'),$$

i.e., S' is at least as good as S^* with respect to both criteria, which proves the result. □

Using Theorem 4.3, (BSL) and its two e-constraint problems can be formulated as integer programs. To keep track of the population covered by the new stops, we have to know which demand points are covered and which not. In addition to the variables x_s defined for (CSL), we therefore define another set of binary variables

$$y_d = \begin{cases} 1 & \text{if demand point } d \text{ is covered} \\ 0 & \text{otherwise.} \end{cases}$$

and let $w = (w_1, w_2, \ldots, w_{|\mathcal{D}|})$.
The IP model of (BSL) can be formulated as

$$\begin{aligned} \min \quad & \begin{pmatrix} cx \\ -wy \end{pmatrix} \\ s.t. \quad & A^{\text{cov}} x \ge y \\ & x \in \{0,1\}^{|\mathcal{S}|} \\ & y \in \{0,1\}^{|\mathcal{D}|}. \end{aligned}$$

The IP model for (BSL-time) is

$$\begin{aligned} \min \quad & cx \\ \text{s.t.} \quad & A^{\text{cov}} x - y \ge 0 \\ & wy \ge Q_{\text{cover}} \\ & x \in \{0,1\}^{|\mathcal{S}|} \\ & y \in \{0,1\}^{|\mathcal{D}|}, \end{aligned} \tag{4.1}$$

and (BSL-cover) is given by

$$\begin{aligned} \max \quad & wy \\ \text{s.t.} \quad & A^{\text{cov}} x - y \ge 0 \\ & cx \le Q_{\text{time}} \\ & x \in \{0,1\}^{|\mathcal{S}|} \\ & y \in \{0,1\}^{|\mathcal{D}|}. \end{aligned} \tag{4.2}$$

In Lemma 3.11 and Theorem 3.13 we pointed out that A^{cov} has the consecutive ones property in the case that the set of tracks $\mathcal{T}$ consists only of a single edge, and under some (weak) assumptions also if $\mathcal{T}$ is a polygonal line. We now discuss variants of (BSL-time) and (BSL-cover), in which

- $w_d = 1$ for all $d \in \mathcal{D}$, $c_e = 1$ for all $e \in E$, and $c_v = 1$ for all $v \in V$, and in which furthermore
- the covering matrix A^{cov} has the consecutive ones property.

Consider the example depicted in Figure 4.1 and note that the coefficient matrix in this example is

$$A^{\text{cov}} = \begin{pmatrix} 1\,1\,1 \\ 0\,1\,1 \end{pmatrix},$$

which has the consecutive ones property.

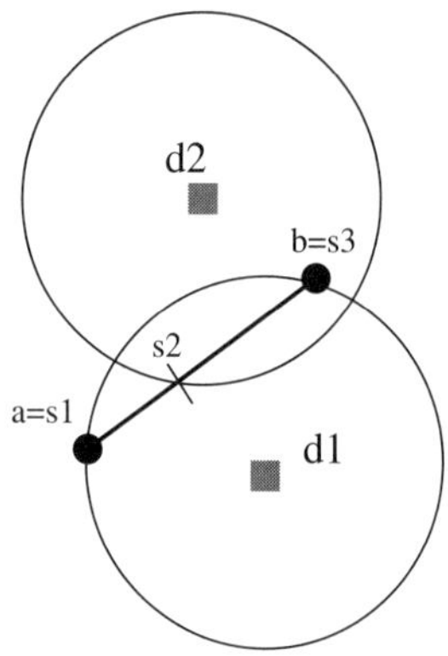

Fig. 4.1. The coefficient matrix of (BSL-time) is not totally unimodular.

(BSL-time): Although A^{cov} has the consecutive ones property that does not yield a totally unimodular coefficient matrix for (BSL-time). Namely, the coefficient matrix of (BSL-time) is in this example given as

$$\begin{pmatrix} 1\,1\,1 & -1 & 0 \\ 0\,1\,1 & 0 & -1 \\ 0\,0\,0 & 1 & 1 \end{pmatrix},$$

which is not totally unimodular.

(BSL-cover): On the other hand, using the same example for (BSL-cover) the coefficient matrix is given by

$$\begin{pmatrix} 1 & 1 & 1 & -1 & 0 \\ 0 & 1 & 1 & 0 & -1 \\ -1 & -1 & -1 & 0 & 0 \end{pmatrix},$$

which is a totally unimodular matrix.

To get rid of the coupling constraint $wy \geq Q_{\text{cover}}$ in (BSL-time) one can consider the Lagrange-relaxation with respect to this constraint.

$$\begin{array}{ll} \min cx + \lambda(Q_{\text{cover}} - wy) & \\ \text{s.t.} \qquad A^{\text{cov}}x - y \geq 0 & \\ \qquad x \in \{0,1\}^{|\mathcal{S}|} & \\ \qquad y \in \{0,1\}^{|\mathcal{D}|}, & \end{array} \tag{4.3}$$

where $\lambda > 0$.

Lemma 4.4. *Let A^{cov} have the consecutive ones property and assume that $w_d = 1$ for all $d \in \mathcal{D}$, and $c_s = 1$ for all $s \in \mathcal{S}$. Then the following hold.*

1. *The Lagrange-relaxation of (BSL-*time*) given in (4.3) (with fixed multiplier λ) can be solved by linear programming.*
2. *(BSL-*cover*) can be solved by linear programming.*

Proof. Since A^{cov} is totally unimodular the matrix $\begin{pmatrix} A^{\text{cov}} & -I \end{pmatrix}$ also is totally unimodular (see Appendix A, or [NW88]). Since this is the coefficient matrix of (4.3), and since the objective function of (4.3) is linear for fixed $\lambda > 0$, this shows the first part of the lemma.
For part 2, we note that $\begin{pmatrix} A^{\text{cov}} \\ 1\ 1\ \ldots\ 1 \end{pmatrix}$ has the consecutive ones property and hence is totally unimodular. Thus, also $\begin{pmatrix} A^{\text{cov}} \\ -1\ -1\ \ldots\ -1 \end{pmatrix}$ is totally unimodular and hence also the coefficient matrix

$$\begin{pmatrix} A^{\text{cov}} & -I \\ -1\ -1\ \ldots\ -1 & 0\ 0\ \ldots\ 0 \end{pmatrix}$$

of (4.2) satisfies this property. □

Note that the lower bound we obtain by solving the Lagrangian dual of (BSL-time) is the same as the lower bound obtained by solving the LP-relaxation of (BSL-time), since Lemma 4.4 shows that all extreme points of the feasible set of (4.3) are integral, see [Wol98]. Further note that for $w_d = 1$ the structure of (BSL-time) resembles the problem (SCP') discussed in Section 3.6, see page 48, such that the solution approaches for (SCP') can be adapted to (BSL-time).

4.3 Bicriteria Set Covering With Consecutive Ones Property

Again, we turn our attention to set covering problems as we already did in Sections 3.4 and 3.6. As before, we do not use any assumptions of the stop location problem, but deal with the bicriteria set covering problem (and its

e-constraint versions) if the coefficient matrix A^{cov} satisfies the consecutive ones property. Let, as before,

$$\mathcal{S} = \{1, 2, \ldots, |\mathcal{S}|\} \text{ and } \mathcal{D} = \{1, 2, \ldots, |\mathcal{D}|\}.$$

The problem we consider is given as

(BSC)

$$\begin{aligned} \min & \begin{pmatrix} cx \\ -wy \end{pmatrix} \\ s.t. \quad & A^{\text{cov}} x \geq y \\ & x \in \{0,1\}^{|\mathcal{S}|} \\ & y \in \{0,1\}^{|\mathcal{D}|}, \end{aligned}$$

where A^{cov} has the consecutive ones property. For $S \subseteq \mathcal{S}$ we define

$$\begin{aligned} x(S)_s &= \begin{cases} 1 & \text{if } s \in S \\ 0 & \text{otherwise} \end{cases} \\ y(S)_d &= \begin{cases} 1 & \text{if } d \in \text{cover}(S) \\ 0 & \text{otherwise} \end{cases} \\ f_{\text{time}}(S) &= cx(S) \\ f_{\text{cover}}(S) &= wy(S). \end{aligned}$$

Since we do not need to cover all rows of A^{cov} in (BSC), we now have to deal with the objective function f_{cover}. To this end, we first investigate cover(S).

Lemma 4.5. *Let $S = \{s_1, \ldots, s_p\} \subseteq \mathcal{S}$ with $s_1 < \cdots < s_p$. Then for all $i = 1, \ldots, p-1$ we have*

$$\text{cover}(s_{i+1}) \setminus \text{cover}\{s_1, \ldots, s_i\} = \text{cover}(s_{i+1}) \setminus \text{cover}(s_i).$$

Proof. Since "$\subseteq$" is trivial, we only need to verify "$\supseteq$". To this end, let

$$d \in \text{cover}(s_{i+1}) \setminus \text{cover}(s_i).$$

We show that $d \notin \text{cover}(s_j)$ for all $j \leq i$. Assume to the contrary that $d \in \text{cover}(s_j)$ and $d \in \text{cover}(s_{i+1})$. This means that $a_{ds_j} = a_{ds_{i+1}} = 1$, and, since A^{cov} has the consecutive ones property also $a_{ds_i} = 1$, a contradiction to $d \notin \text{cover}(s_i)$. □

Lemma 4.5 motivates a dynamic programming approach, which we will develop in the following. To this end we define an acyclic digraph G_{BSC}. Note that we cannot utilize the digraph G_{SC} (see Notation 3.22 on page 37) for solving (BSC), since we now have to allow solutions which do not cover all rows. This implies that we have to include all edges (i, j), if $i < j$, in the edge set of G_{BSC}.

Notation 4.6. *The* **bicriteria set covering digraph** $G_{BSC} = (V_{SC}, E_{BSC})$ *is defined by*

$$V_{BSC} = \mathcal{S} \cup \{s, t\}, \quad \textit{and}$$
$$E_{BSC} = \{(i,j) : i, j \in \mathcal{S} \textit{ and } i < j\} \cup \{(s,j) : j \in \mathcal{S}\} \cup \{(i,t) : i \in \mathcal{S}\} \cup \{(s,t)\}.$$

For each edge $(i,j) \in E_{BSC}$ *we furthermore define costs and weights*

$$c_{ij} = \begin{cases} c_j & \textit{if } j \neq t \\ 0 & \textit{if } j = t \end{cases}$$
$$w_{ij} = \begin{cases} \sum_{d \in \text{cover}(j) \setminus \text{cover}(i)} w_d & \textit{if } i \neq s, j \neq t \\ \sum_{d \in \text{cover}(j)} w_d & \textit{if } i = s, j \neq t \\ 0 & \textit{if } j = t. \end{cases}$$

Moreover, for an s-t-path S *in* G_{BSC} *let* $C(S)$ *denote its length according to* c_{ij} *and* $W(S)$ *denote its length according to* w_{ij}.

As in G_{SC} (see Corollary 3.24 on page 39) any $S \subseteq \mathcal{S}$ uniquely defines an s-t-path in G_{BSC} by adding the nodes s and t to S. Moreover,

$$C(S \cup \{s,t\}) = f_{\text{time}}(S).$$

In the next result we state that the same holds for the objective function f_{cover}. Note that for proving this observation we need that the coefficient matrix A^{cov} has the consecutive ones property.

Theorem 4.7. *Let* $S \subseteq \mathcal{S}$. *Then* $W(S \cup \{s,t\}) = f_{\text{cover}}(S)$.

Proof. We use induction on $p = |S|$. For $p = 1$ the claim is true due to the definition of $w_{sj} = f_{\text{cover}}(\{j\})$ and $w_{jt} = 0$. Now assume that

$$W(S' \cup \{s,t\}) = f_{\text{cover}}(S')$$

for all S' with $|S'| \leq p$. Take some $S = \{s_1, s_2, \ldots, s_p, s_{p+1}\}$ and assume that $s_1 < s_2 < \cdots < s_{p+1}$. Define $S' = \{s_1, s_2, \ldots, s_p\}$. Then we get

$$\begin{aligned} f_{\text{cover}}(S) &= \sum_{d \in \text{cover}(S)} w_d \\ &= \sum_{d \in \text{cover}(S')} w_d + \sum_{d \in \text{cover}(s_{p+1}) \setminus \text{cover}(S')} w_d \\ &= W(S' \cup \{s,t\}) + \sum_{d \in \text{cover}(s_{p+1}) \setminus \text{cover}(s_p)} w_d \qquad (4.4) \\ &= W(S \cup \{s,t\}), \end{aligned}$$

where (4.4) holds due to Lemma 4.5. □

As an example, consider again the matrix

$$A^{\text{cov}} = \begin{pmatrix} 1\,1\,1\,0\,0\,0 \\ 0\,1\,1\,1\,0\,0 \\ 0\,0\,1\,1\,1\,0 \\ 0\,0\,0\,0\,1\,1 \end{pmatrix}$$

and let

$$\begin{aligned} c_1 &= 3 \quad w_1 = 3 \\ c_2 &= 1 \quad w_2 = 2 \\ c_3 &= 2 \quad w_3 = 3 \\ c_4 &= 1 \quad w_4 = 1 \\ c_5 &= 3 \\ c_6 &= 2. \end{aligned}$$

Let us investigate the path $P = \{s, 2, 3, 4, t\}$ depicted in Figure 4.2. First, we determine

$$\begin{aligned} c_{s2} &= 1 \quad w_{s2} = 5 \\ c_{23} &= 2 \quad w_{23} = 3 \quad \text{since cover}(3) \setminus \text{cover}(2) = \{3\} \\ c_{34} &= 1 \quad w_{34} = 0 \quad \text{since cover}(4) \setminus \text{cover}(3) = \emptyset \\ c_{4t} &= 0 \quad w_{4t} = 0. \end{aligned}$$

This yields $C(P) = 1 + 2 + 1 = 4$ and $W(P) = 5 + 3 = 8$. On the other hand, for the corresponding set $S = P \setminus \{s, t\} = \{2, 3, 4\}$ we obtain

$$\begin{aligned} f_{\text{time}}(S) &= 1 + 2 + 1 = 4 \\ \text{cover}(S) &= \{1, 2, 3\}, \text{ and hence} \\ f_{\text{cover}}(S) &= 3 + 2 + 3 = 8, \end{aligned}$$

such that $W(P) = f_{\text{cover}}(S)$ and $C(P) = f_{\text{time}}(S)$ holds.

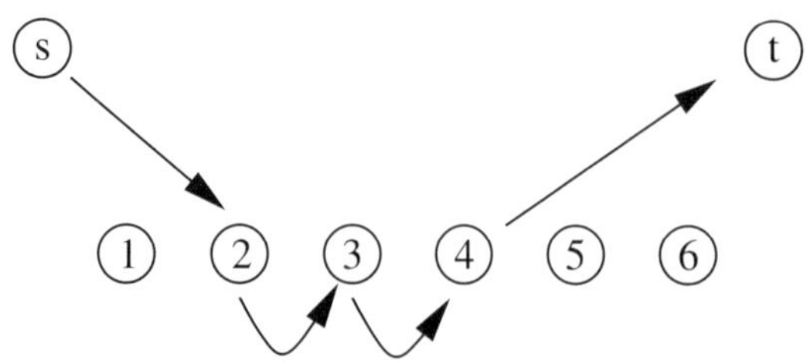

Fig. 4.2. The path S in the digraph G_{BSC} for the example.

Corollary 4.8. *Let $S \subseteq \mathcal{S}$. Then S is a Pareto solution of (BSC) if and only if $S \cup \{s, t\}$ is a path in G_{BSC} which is Pareto minimal with respect to the two length functions C and W.*

This means that finding all efficient solutions of (BSC) reduces to finding all efficient paths in the acyclic network G_{BSC}. Note further that using negative weights $-w(e)$ does not affect the solution procedures or the complexity of the problem, since no directed cycles exist. Finding efficient paths in the presence of more than one objective function belongs to the most widely studied multiobjective combinatorial optimization problems. We refer to the surveys given in Section 6.1. of [EG02] and in [Skr00], where the latter deals in particular with bicriteria shortest path problems. Available algorithms are based on dynamic programming (as presented in [Hen85]), or on label setting methods (see [Han79] for an early contribution). Label correcting methods for bicriteria shortest path problems were presented, e.g., in [MMO91, SA00]. An algorithm based on a ranking of paths was proposed by [CM82].

Finally, we turn our attention to the special case in which $c_s = 1$ for all $s \in \mathcal{S}$. In this case, $f_{\text{time}}(S) = |S|$ is equal to the number of edges of the path $S \cup \{s, t\}$, reduced by 1. Letting $Q_{\text{time}} = K \in \mathbb{N}$, the cost-constraint version of (BSC) is given as

(BSC-cover**(K))**

$$\begin{aligned} \max \quad & wy \\ \text{s.t. } A^{\text{cov}}x - y &\geq 0 \\ cx &\leq K \\ x &\in \{0,1\}^{|\mathcal{S}|} \\ y &\in \{0,1\}^{|\mathcal{D}|}. \end{aligned} \tag{4.5}$$

This problem can be solved by finding a longest path with no more than $K+1$ edges, i.e., by solving a cardinality constraint shortest path problem in an acyclic digraph. This can be done by the Algorithm of Bellman–Ford (see, [Bel58, FF62]), which needs $O(K|\mathcal{S}|^2)$ time in the worst case to find a longest path with no more than K edges from one specified starting node to all other nodes in the graph. To find all efficient solutions we again use the result of [HC83].

Lemma 4.9. *Let K^* be the solution of (SCP), i.e., the minimal number of columns needed to cover all rows of A^{cov}, and let $K \leq K^*$. Then, S is an optimal solution of (BSC-*cover*(K)) if and only if S is a Pareto solution.*

Proof. If S is the unique solution of (BSC-cover(K)) the result follows directly from [HC83] (see also Lemma 4.1). Now assume that $S_1 \neq S_2$ but both are optimal solutions of (BSC-cover(K)). Then $f_{\text{cover}}(S_1) = f_{\text{cover}}(S_2)$, $f_{\text{time}}(S_1) \leq K$, and $f_{\text{time}}(S_2) \leq K$. We want to show that

$$f_{\text{time}}(S_1) = f_{\text{time}}(S_2),$$

since in this case both solutions lead to the same efficient point and no solution dominating S_1 and S_2 in both criteria exists.
Assume to the contrary that $|S_1| = f_{\text{time}}(S_1) < f_{\text{time}}(S_2) = |S_2|$. In particular, $|S_1| < |S_2| \leq K \leq K^*$, i.e., $\mathcal{D} \setminus \text{cover}(S_1) \neq \emptyset$. But this means that there exists a column $s \in \mathcal{S} \setminus S_1$ such that adding s to S_1 would increase f_{cover}, i.e.,

$$f_{\text{cover}}(S_1 \cup \{s\}) > f_{\text{cover}}(S_1) = f_{\text{cover}}(S_2)$$

and $f_{\text{time}}(S_1 \cup \{s\}) \leq K$, which is a contradiction to the optimality of S_1 and S_2 for (BSC-cover(K)). □

In the following algorithm we make use of this result. If the assumptions of Theorem 3.13 are satisfied we determine all efficient solutions of (BSC) by solving a sequence of cardinality constraint longest path problems using the algorithm of Bellman–Ford. Since one run of the algorithm for K^* determines also all solutions with cardinality constraints given by smaller K, i.e., for all $K \in \{0, 1, \ldots, K^*\}$ the overall complexity of the following algorithm equals the complexity of the algorithm of Bellmann–Ford for one single source node s, i.e., we obtain a complexity of $O(|\mathcal{S}|^3)$ for the determination of all efficient points of (BSC). For our special case of the stop location problem recall further that for each demand point d we have at most two candidates in the finite dominating set $\mathcal{S}$ such that the complexity of finding all efficient solutions in the case of a matrix A^{cov} with consecutive ones property, traffic loads $c_e = c_v = 1$ and arbitrary weights w_d is bounded by $O(|\mathcal{D}|^3)$.

Algorithm 10: Finding all efficient points for (BSC) with consecutive ones property and $c_s = 1$

Input: w, A^{cov} with consecutive ones property, $c_s = 1$ for all $s \in \mathcal{S}$.
Output: All efficient points for (BSC), and a Pareto solution for each of them.
Step 1. Use Algorithm 2 to transform A^{cov} into a strictly monotone matrix.
Step 2. Solve the unweighted set covering problem by Algorithm 3, let K^* be the cardinality of the optimal solution.
Step 3. Construct G_{BSC}.
Step 4. Use the algorithm of Bellmann-Ford to find all longest paths w.r.t. the weights w from s to t with $K = 1, 2, \ldots K^*$ edges. Let h^K denote the length of a longest s-t-path P^K with at most K edges.
Step 5. Output: $Eff = \{(h^K, K) : K = 1, \ldots, K^*\}$ with corresponding Pareto solutions P^K, $K = 1, \ldots, K^*$.

4.4 Varying the Radius

The finite dominating set $\mathcal{S}$ depends on the given covering radius r. If more than one value for r should be considered it would be advantageous to have a finite dominating set which is **not** dependent on the specific value of r. Such a set will be derived in this section, for the special case that $c_v \leq c_e$ for all $v \in V$ and for all edges $e \in E$ incident with v.
Using the same denotation as on page 26 we define

$$\mathcal{S}^* = \{s \in \mathcal{T} : \text{ there exist } d_1, d_2 \in \mathcal{D}, d_1 \neq d_2 \text{ such that } \gamma_{d_1}(d_1, s) = \gamma_{d_2}(d_2, s)\}$$
$$\cup \bigcup_{d \in \mathcal{D}, e \in E} \text{argmin}_{s \in e} \gamma_d(d, s) \cup \bigcup_{e \in E} \left\{ \frac{v_1^e + v_2^e}{2} \right\} \cup V,$$

where v_1^e, v_2^e are the two endpoints of the edge $e \in E$. The first set of candidates can be determined by intersecting the bisector between each pair of demand points with $\mathcal{T}$, the second set describes the projection points from all demand points onto all edges, and the third set makes sure that V and one point of the interior of each edge is included in $\mathcal{S}^*$.

Theorem 4.10. *$\mathcal{S}^*$ is a dominating set for (BSL), (BSL-*time*), and (BSL-*cover*) for any given radius $r > 0$, if $c_v \leq c_e$ for all $v \in V$, $e \in E$ if e is incident with v.*

Proof. Take $r > 0$ arbitrary, but fixed. Using Theorems 3.7 and 4.3 we can assume that either the problems are infeasible, or there exists an optimal solution $S^* \subseteq \mathcal{S}$, where $\mathcal{S}$ is the specific candidate set with respect to r as defined in (3.2) on page 27. Take $s \in S^*$ with $s \notin \mathcal{S}^*$. Hence, $s \not\subset V$ and $s \neq \frac{v_1^e + v_2^e}{2}$ for any of the edges $e \in E$. Consequently, s lies in the interior of some edge e, and there exists exactly one $d \in \mathcal{D}$ such that

$$\gamma_d(d, s) = r$$

and

$$\text{argmin}_{s' \in e} \gamma_d(d, s') \neq s,$$

otherwise, $s \in \mathcal{S}^*$. Move s along e within B_d^r (i.e., get closer to d) to a new point, until one of the following conditions is satisfied for the first time.

- s reaches some point in $\mathcal{S}^*$, let this point be denoted by s^*.
- s leaves some ball $B_{d'}^r$: Denote this point by s'. It is important to note that $d \neq d'$, since otherwise the projection point from d onto e would have been reached before. Then we have

$$\gamma_{d'}(d', s') = r \text{ and } \gamma_d(d, s') \leq r$$
$$\gamma_{d'}(d', s) \leq r \text{ and } \gamma_d(d, s) = r,$$

  meaning that the continuous function $h : e \to \mathbb{R}$ defined by $h(x) = \gamma_{d'}(d', x) - \gamma_d(d, x)$ satisfies $h(s') \geq 0$ and $h(s) \leq 0$ such that the intermediate value theorem yields the existence of a point s^* between s and s' with $h(s^*) = 0$, i.e., $s^* \in \mathcal{S}^*$.

Due to the construction, in both cases we have found a point $s^* \in \mathcal{S}^*$ such that

$$\text{cover}(s) \subseteq \text{cover}(s^*), \text{ and}$$
$$c_{g(s)} \geq c_{g(s^*)},$$

the latter holding since either $g(s) = g(s^*) = e$, or $g(s) = e$ and $g(s^*) = v$ with v and e incident to each other, and hence satisfying $c_{g(s)} = c_e \geq c_v = c_{g(s^*)}$. Finally we get that $S' = S^* \setminus \{s\} \cup \{s^*\}$ satisfies

$$f_{\text{cover}}(S^*) \leq f_{\text{cover}}(S')$$
$$f_{\text{time}}(S^*) \geq f_{\text{time}}(S').$$

Since the above argument can be repeated for all $s \in S^* \setminus \mathcal{S}^*$ this completes the proof. □

We remark that, dependent on the structure of the bisectors, $\mathcal{S}^*$ needs not be a finite set (for a detailed discussion on determining bisectors for gauge distance functions, see [Wei99]), but can be made finite along the lines of Chapter 3.2. We further point out that the size of $\mathcal{S}^*$ is $O(|\mathcal{D}|^2 + |\mathcal{D}||E|)$, which is much larger than the size of $\mathcal{S}$, which is of order $O(|\mathcal{D}|)$. Nevertheless, $\mathcal{S}^*$ also is a finite candidate set for problem (CSL) discussed in Chapter 3.

While the assumption of Theorem 4.10 is in practice usually satisfied for all nodes with degree at most 2, it may happen that it does not hold for nodes with more than two incident edges. Unfortunately, the next result implies that no finite dominating set is possible without this assumption.

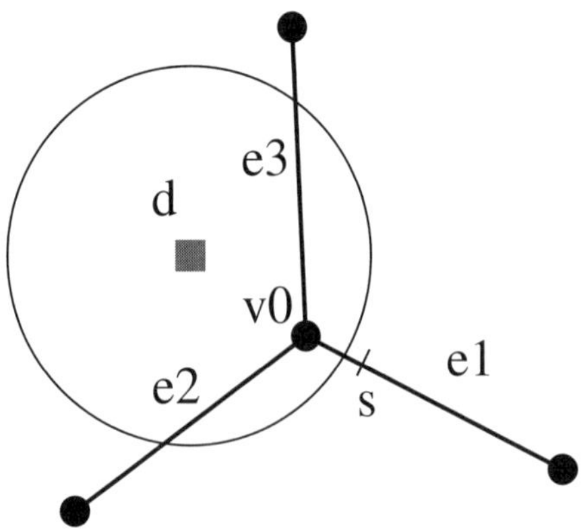

Fig. 4.3. No finite dominating set is possible independent of r.

Lemma 4.11. *Without the assumptions of Theorem 4.10 no finite dominating set which can be used for any value of r exists for (BSL), (BSL-*time*), and for (BSL-*cover*).*

Proof. Consider a star-shaped graph with four nodes and three edges, all edges incident with node v_0 (see Figure 4.3). Let one demand point d be given, such that $\operatorname{argmin}_{s \in e_1} \gamma_d(d,s) = v_0$, and consider, e.g., the Euclidean distance l_2. Let $c_{e_1} = 2, c_{e_2} = c_{e_3} = 100$, and $w_{v_0} = 102$ be (realistic) traffic loads. Now take any finite candidate set $\mathcal{S}^f$, and let s be the candidate of e_1 which is closest to v_0. Note that $l_2(d,v_0) < l_2(d,s)$. Moreover, let r be such that

$$l_2(d,v_0) < r < l_2(d,s).$$

Then the optimal solution

- of (BSL-time) with any $Q_{\text{cover}} > 0$, and
- of (BSL-cover) with $Q_{\text{time}} = 1$

is a point on e_1 between v_0 and s, which is not contained in the finite candidate set $\mathcal{S}^f$. Furthermore, such a point dominates any solution $S \subseteq \mathcal{S}^f$, hence $\mathcal{S}^f$ cannot be used to derive all efficient points of (BSL). □

5

Extensions

The extensions we are going to discuss in this chapter have been brought up by our practical studies. First, we relax the requirement that $\mathcal{D}$ consists of points and allow demand regions D instead. This idea is motivated by the accurate data we have for the demand regions in the stop location problem, given as polygons (representing the settlements). The problem of minimizing the additional travel time caused by the new stops, while covering all demand regions, is denoted by (CSL-region), while (BSL-region) refers to the bicriteria variant corresponding to (BSL), if we consider demand regions instead of demand points.

Secondly, we consider alternative objective functions. Instead of looking at the bicriteria variant, we can also sum up the positive and the negative effects of new stops, where we have to measure both effects using the same scale, e.g., minutes of travel time. A new stop decreases the *access time* for customers living close to it, but increases the *travel time* for customers just traveling through this new stop. Summing up we get an approximation of the change in *door-to-door travel time* over all customers.

Chapter 5 is structured as follows: We first reformulate the stop location problem in the case of demand regions instead of demand points. If a finite candidate set is given, we present integer programming formulations. Then we turn our attention back to the continuous case and in particular to (CSL-region). We show that finiteness of the solution is not guaranteed in this case. Nevertheless, we present an efficient (exact) algorithm solving (CSL-region) in the unweighted case. This algorithm is an extension of Algorithm 4.

Then we present a model for minimizing the sum of all changes in the door-to-door travel time over all customers. We show its NP-completeness and briefly describe our experience solving this problem with a genetic algorithm. Other possible objective functions (from the point of view of the public transportation company) are mentioned.

5.1 Covering Demand Regions

In this section we consider a better approximation of the demand set based on using demand regions instead of demand points, see also [SS03]. According to Section 2.3, especially Definition 2.5, we use the following notation. Let $\mathcal{D} = \{D_1, \ldots, D_{|\mathcal{D}|}\}$ be a finite set of connected, and pairwise disjoint demand regions $D_i \subseteq \mathbb{R}^2$, $i = 1, \ldots, |\mathcal{D}|$, and let

$$D_{total} = \bigcup_{D \in \mathcal{D}} D$$

be the set of all points in the demand regions. For each demand region we assume that a distance measure γ_D has been specified as a norm or a gauge. Hence, given some possible new stop $s \in \mathcal{T}$ and a point $d \in D \in \mathcal{D}$, the distance from d to s is given as

$$\gamma_d(d, s) = \gamma_D(d, s).$$

Furthermore, recall that for each $d \in D \in D_{total}$ and each $S \subseteq \mathcal{T}$ we defined

$$\begin{aligned}
\mathcal{T}(d) &= \{s \in \mathcal{T} : \gamma_d(d, s) \leq r\} \\
&= \{s \in \mathcal{T} : d \in \text{cover}(s)\} \quad \text{and} \\
\text{cover}(S) &= \{d \in \mathcal{D} : \gamma_d(d, s) \leq r\} \\
&= \{d \in \mathcal{D} : S \cap \mathcal{T}(d) \neq \emptyset.\}
\end{aligned}$$

Denoting by $\lambda(A)$ the area of a (measurable) set $A \subseteq \mathbb{R}^2$, by w_D the number of (potential) customers of demand region D, and by $c_{g(s)}$ the traffic load of the edge or of the node where s is located, the two objective functions f_{cover} and f_{time} for demand regions are given by

$$\begin{aligned}
f_{\text{cover}}(S) &= \sum_{D \in \mathcal{D}} w_D \frac{\lambda(\text{cover}(S) \cap D)}{\lambda(D)} \\
f_{\text{time}}(S) &= \sum_{s \in S} c_{g(s)}.
\end{aligned}$$

(CSL-region) is hence given as follows.

(CSL-region)

Given $G = (V, E)$ with its set of points $\mathcal{T} = \bigcup_{e \in E} e$, traffic loads c_e for all $e \in E$, and c_v for all $v \in V$, as well as a finite set of demand regions $\mathcal{D}$, with associated gauges γ_D for all $D \in \mathcal{D}$, find a set $S \subseteq \mathcal{T}$ such that $\text{cover}(S) = \mathcal{D}$ *and*

$$f_{\text{time}}(S) = \sum_{s \in S} c_{g(s)}$$

is minimized.

Similarly, the bicriteria stop location problem in case of demand regions instead of demand points can be restated as follows.

(BSL-region)

Given $G = (V, E)$ with its set of points $\mathcal{T}$, with traffic loads c_e for all $e \in E$, and c_v for all $v \in V$, as well as a finite set of demand regions $\mathcal{D}$ with weights w_D and gauges γ_D for all $D \in \mathcal{D}$, find a set $S \subseteq \mathcal{T}$ such that both

$$f_{\text{time}} = \sum_{s \in S} c_{g(s)}, \quad \textit{and}$$

$$-f_{\text{cover}}(S) = -\sum_{D \in \mathcal{D}} w_D \frac{\lambda(\text{cover}(S) \cap D)}{\lambda(D)}$$

are minimized.

Note that—in contrast to (CSL) and (BSL) for demand points—(CSL-region) is **not** a special case of the cover-constraint version of (BSL-region), since

$$\frac{\lambda(D \cap \text{cover}(S))}{\lambda(D)} = 1$$

does **not** imply $D \subseteq \text{cover}(S)$. We first state the NP hardness of both problems.

Theorem 5.1.

1. *(CSL-region) is NP-hard.*
2. *(BSL-region) is NP-hard.*

Proof.

1. If $\mathcal{D}$ is a set of points it is a special case of (CSL-region), thus the NP-hardness of (CSL-region) follows from the NP-hardness of (CSL), see Theorem 3.2.
2. We can reduce (BSL-region) to (CSL) as follows. Let $\mathcal{D}$ be a finite set of demand points, and let $K \geq 0$. We know that it is NP-hard to decide if $\mathcal{D}$ can be covered by K discs all with center points in $\mathcal{T}$, see Theorem 3.2. Now enlarge each demand point to a small disc B_d^ϵ centered at d. Increasing r to $r + \epsilon$ yields an instance of (BSL-region)
   - with $\mathcal{D}' = \{B_d^\epsilon : d \in \mathcal{D}\}$,
   - with γ_D is the Euclidean distance for all $D \in \mathcal{D}'$,
   - with $w_{B_d^\epsilon} = w_d$, and
   - with $Q_{\text{time}} = K$ and $Q_{\text{cover}} = \sum_{D \in \mathcal{D}} w_D$.

   This instance satisfies the following:
   $\mathcal{D}$ can be covered by at most K discs of radius r with center points in $\mathcal{T}$, if and only if all $B_d^\epsilon \in \mathcal{D}'$ can be covered by at most K discs of radius $r + \epsilon$ which all have their center points in $\mathcal{T}$, i.e., if and only if there exists $S \subseteq \mathcal{T}$ with $f_{\text{time}}(S) \leq Q_{\text{time}}$ and $f_{\text{cover}}(S) \geq Q_{\text{cover}}$. □

The Discrete Stop Location Problem for Demand Regions

Let us first assume that a discrete set of candidates $\mathcal{S}^{cand} \subseteq \mathcal{T}$ is given, and that we have to find a set of new stops $S \subseteq \mathcal{S}^{cand}$. Then (CSL-region) and (BSL-region) can be formulated as integer programs.

Notation 5.2. *Let x, y be two points in D_{total}.*

$$x \sim y \text{ if } \left(\forall s \in \mathcal{S}^{cand} : x \in \text{cover}(s) \Longleftrightarrow y \in \text{cover}(s)\right).$$

Note that $\sim$ is an equivalence relation. Hence we obtain a partition of D_{total} into equivalence classes. We define a set of cells $\mathcal{C}$ by intersecting the equivalence classes with all $D \in \mathcal{D}$. Due to the construction of $\mathcal{C}$ we obtain the following lemma.

Lemma 5.3. *Let $C \in \mathcal{C}$.*

1. *If $C \cap \text{cover}(s) \neq \emptyset$ for some $s \in \mathcal{S}^{cand}$ then $C \subseteq \text{cover}(s)$.*
2. *If $C \cap D \neq \emptyset$ for some $D \in \mathcal{D}$ then $C \subseteq D$.*

Now, define

$$x_s = \begin{cases} 1 & \text{if } s \in \mathcal{S}^{cand} \text{ is chosen as new stop} \\ 0 & \text{otherwise.} \end{cases}$$

and let the following matrix $A^{\text{cov}} = (a_{Cs})_{C \in \mathcal{C}, s \in \mathcal{S}^{cand}}$ contain the covering information.

$$a_{cs} = \begin{cases} 1 & \text{if } C \subseteq \text{cover}(s) \\ 0 & \text{otherwise} \end{cases}$$

First we give the integer programming formulation for covering all demand regions with a minimal cost set of stops, i.e., for the problem

(Discrete-CSL-region)

$$\begin{aligned} &\min f_{\text{time}}(S) \\ &\text{s.t. } \text{cover}(S) = D_{total} \\ &\qquad\qquad\quad S \subseteq \mathcal{S}^{cand}. \end{aligned}$$

To obtain an integer programming formulation we reformulate (Discrete-CSL-region) as follows.

$$\begin{aligned} \min \quad & cx \\ \text{s.t. } & A^{\text{cov}} x \geq \mathbf{1}_{|\mathcal{C}|} \\ & x \in \{0,1\}^{|\mathcal{S}^{cand}|}. \end{aligned} \tag{5.1}$$

For the bicriteria formulation we need to calculate $\lambda(C)$ for all cells $C \in \mathcal{C}$ and save these values in the vector $\lambda \in \mathbb{R}^{|\mathcal{C}|}$. For $C \subseteq D$ we determine

$$w_C = w_D \frac{\lambda_C}{\lambda_D}.$$

Furthermore we define

$$y_C = \begin{cases} 1 & \text{if cell } C \text{ is covered} \\ 0 & \text{otherwise} \end{cases}$$

such that we can formulate

(Discrete-BSL-region)

$$\begin{array}{ll} \min & \begin{pmatrix} f_{\text{time}}(S) \\ -f_{\text{cover}}(S) \end{pmatrix} \\ \text{s.t.} & S \subseteq \mathcal{S}^{cand} \end{array}$$

as the following IP model:

$$\begin{array}{ll} \min & \begin{pmatrix} cx \\ -wy \end{pmatrix} \\ s.t. & A^{\text{cov}} x \geq y \\ & x \in \{0,1\}^{|\mathcal{S}^{cand}|} \\ & y \in \{0,1\}^{|\mathcal{C}|}. \end{array}$$

Note that these formulations are again set covering formulations. Hence, if A^{cov} has the consecutive ones property we can apply the solution methods developed in Sections 3.4 to solve (Discrete-CSL-region) and the solution methods of Section 4.3 for (Discrete-BSL-region). If A^{cov} almost has the consecutive ones property, we can apply the methods of Section 3.6.

Covering All Demand Regions

We turn our attention back to (CSL-region) in the continuous case, i.e., we want to find a set of stops in $\mathcal{T}$ with minimal cost such that all demand regions are covered. Unfortunately, (CSL-region) does not have the nice properties of the corresponding complete cover problem in the case of demand points. One difference concerns the feasibility of the problem.

Lemma 5.4.

1. *(CSL-region) is feasible if and only if* $D_{total} \subseteq \text{cover}(\mathcal{T})$.
2. *(CSL-region) may be feasible but have no finite solution.*

Proof.

1. If $D_{total} \subseteq \text{cover}(\mathcal{T})$ then $\mathcal{T}$ is a feasible solution, otherwise there exists $d \in D_{total}$ that can not be covered by any point in $\mathcal{T}$.
2. An example of a problem for which no finite feasible solution exists is given in Figure 5.1. □

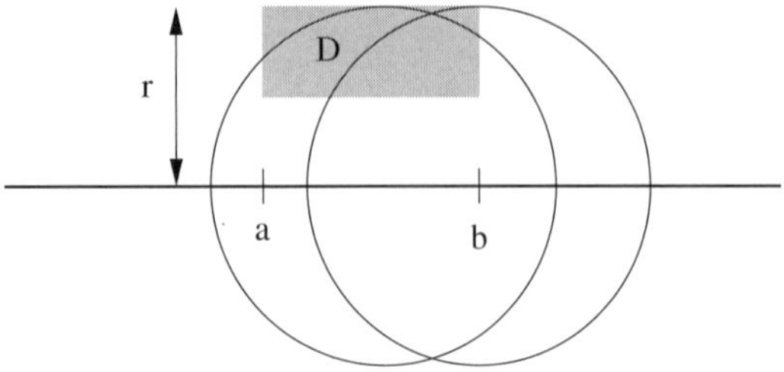

Fig. 5.1. The line segment between a and b is a solution of (CSL-region), but no finite solution exists.

For the remainder of this section we assume that (CSL-region) is feasible, i.e.,

$$D_{total} \subseteq \text{cover}(\mathcal{T}).$$

Unfortunately it is not possible to transfer the finite candidate set from (SCP) by using the intervals $\mathcal{T}(d)$ as in the case of demand points. Intuitively, one could determine $\mathcal{T}(D)$ as the set of all points which can be used to cover D, i.e.,

$$\mathcal{T}(D) = \{s \in \mathcal{T} : D \subseteq \text{cover}(s)\},$$

and take the endpoints of these intervals $\mathcal{T}(D)$ as a candidate set. But this candidate set need not contain the optimal solution, it even need not contain any feasible solution (also in the case that a finite feasible solution exists). Such situations are shown in Figures 5.2 and 5.3 for the case that γ_D is the Euclidean distance for all $D \in \mathcal{D}$.

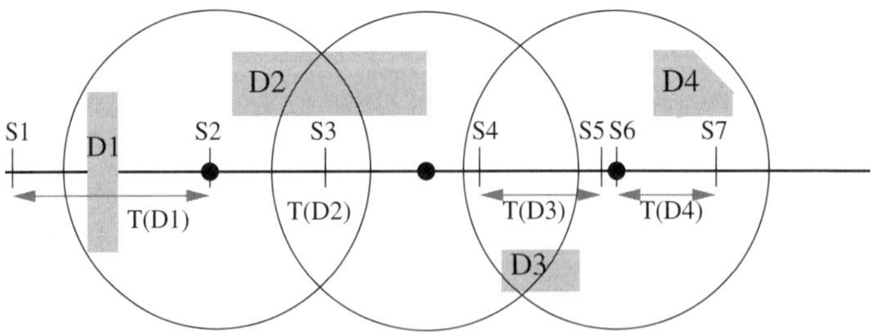

Fig. 5.2. The set of endpoints of the intervals $\mathcal{T}(D)$ and an optimal solution.

- In Figure 5.2 we have four demand regions D_1, D_2, D_3, D_4. For some radius r we determine

$$\mathcal{T}(D_1) = [s_1, s_2]$$
$$\mathcal{T}(D_2) = \{s_3\}$$
$$\mathcal{T}(D_3) = [s_4, s_5]$$
$$\mathcal{T}(D_4) = [s_6, s_7]$$

such that the candidate set consisting of the endpoints of these intervals would be

$$\{s_1, s_2, s_3, s_4, s_5, s_6, s_7\}.$$

Within this candidate set the best possible solution contains four stops, e.g., $\{s_2, s_3, s_4, s_6\}$. But the optimal solution of this problem instance only needs three new stations, namely the thick points in Figure 5.2, and only two of them are endpoints of intervals $\mathcal{T}(D)$.

- The situation is even worse in the case of Figure 5.3. Here a demand region D is depicted which is too large to be covered by only one stop, i.e.,

$$\mathcal{T}(D) = \emptyset,$$

leading to an empty candidate set, although the problem is (finitely) feasible with only two new stops, see the thick points in Figure 5.3.

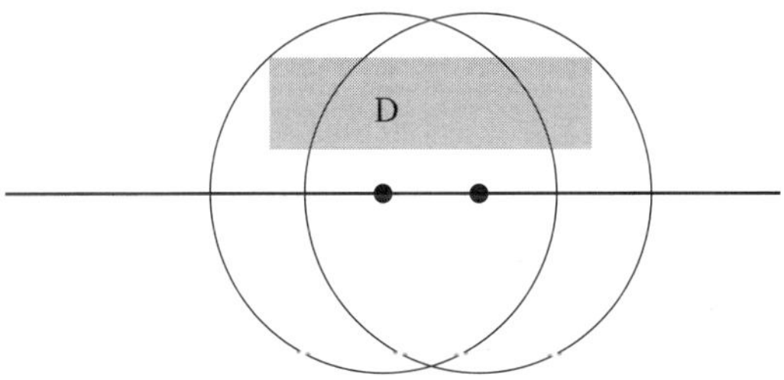

Fig. 5.3. The set of endpoints of the intervals $\mathcal{T}(D)$ is empty, although the problem is feasible.

Nevertheless, we now develop an algorithm for solving (CSL-region) in the continuous case. This algorithm finds the optimal solution on a straight line, or on a polygonal line satisfying the assumptions of Theorem 3.13, if the costs c_e of all edges and c_v of all nodes belonging to the polygonal line are equal. Recall that for $d \in D_{total}$

$$\mathcal{T}(d) = \{s \in \mathcal{T} : \gamma_d(d, s) \leq r\}$$

is that part of the tracks $\mathcal{T}$ which can be used to cover d, i.e., we know that at least one point of $\mathcal{T}(d)$ has to be chosen as a new stop for each $d \in D_{total}$. Furthermore we know that for an edge e with $\mathcal{T}(d) \cap e \neq \emptyset$ there exist two (not necessarily distinct) points $f_d^e, l_d^e \in e$ such that

$$\mathcal{T}(d) \cap e = [f_d^e, l_d^e],$$

see Lemma 3.4 and the discussion on page 25. Moreover, recall from Lemma 3.9 that

$$d \in \text{cover}(s) \Longleftrightarrow s \in \mathcal{T}(d).$$

For $\mathcal{T}(d) \subseteq e$ we hence obtain

$$d \in \text{cover}(s) \Longleftrightarrow f_d^e \leq s \leq l_d^e. \tag{5.2}$$

Since all $\mathcal{T}(d)$ are intervals we can write

$$\mathcal{T}(d) = [f_d, l_d].$$

We use this notation for our next definition.

Definition 5.5. *Let $\mathcal{T} = e$. Given some set $Q \subseteq D_{total}$ with $Q \subseteq \text{cover}(e)$, the (left)* **fixturing point** *of Q is*

$$L(Q) = \inf\{l_d : d \in Q\} \in e.$$

We are now in the position to present our algorithm for solving (CSL-region). Note that we assumed that the costs for all $s \in \mathcal{T}$ are the same, such that our goal is to find the minimal number of stops needed to cover D_{total}.

Algorithm 11: Solving (CSL-region) along a single edge

```
Input: T = e, D ⊆ cover(T), γ_D for all D ∈ D.
Output: Optimal solution S* of (CSL-region).
Step 1. Set S* = ∅, D = D_total.
Step 2. S* = S* ∪ {L(D)}, D = D \ cover(L(D)).
Step 3. If D = ∅, stop, otherwise goto 2.
```

Note that the algorithm need not terminate finitely. But, fortunately, in Theorem 5.7 we can show that this only happens in the case that no finite optimal solution exists. Moreover, we will show the correctness of the algorithm, i.e., that in case of finite termination of the algorithm, we obtain the optimal solution of (CSL-region) on a single edge.

In the following, let y_i denote the fixturing point found in iteration i and

$$D_i = D_{total} \setminus \text{cover}\{y_1, \ldots, y_{i-1}\}$$

be the set D at the beginning of iteration i, i.e.,

$$y_i = L(D_i).$$

We first state the following lemma.

Lemma 5.6.

1. *$y_i \leq y_{i+1}$ for $i = 1, 2, \ldots$.*
2. *If the algorithm terminates, then $y_i < y_{i+1}$ for $i = 1, 2, \ldots$.*
3. *Let Y^* be the output of the algorithm in case that it terminates. Then Y^* is a feasible solution of (CSL-region).*

Proof.

1. Since $D_{i+1} = D_i \setminus \text{cover}(y_i) \subseteq D_i$ we obtain
$$\begin{aligned} y_i &= L(D_i) = \inf\{l_d : d \in D_i\} \\ &\leq \inf\{l_d : d \in D_{i+1}\} \\ &= L(D_{i+1}) = y_{i+1}. \end{aligned}$$
2. Suppose $D_i \neq \emptyset$, but the fixturing point found in iteration i satisfies $y_{i-1} = y_i$. Then
$$\begin{aligned} D_{i+1} &= D_i \setminus \text{cover}(y_i) \\ &= D_i \setminus \text{cover}(y_{i-1}) = D_i \neq \emptyset, \end{aligned}$$
and hence $D_k = D_i$ for all $k \geq i$, i.e., the algorithm does not terminate.
3. Let Y^* be the output of the algorithm. Consider $d \in D_{total}$. We want to show that $d \in \text{cover}(Y^*)$. Since the algorithm has terminated finitely, say at the end of iteration I, we know that $D_{I+1} = \emptyset$. In particular, $d \notin D_{I+1}$. Let $d \in D_i$, but $d \notin D_{i+1}$. Then $d \in \text{cover}(L(D_i))$, i.e., $d \in \text{cover}(Y^*)$. □

Based on the results above, we now present the proof for the correctness of Algorithm 11.

Theorem 5.7. *If there exists a finite solution of (CSL-region) then Algorithm 11 terminates with an optimal solution Y^*.*

Proof. Let $Y^* = \{y_1, y_2, \ldots\}$ be the set generated by Algorithm 11. According to part 1 of Lemma 5.6 we know that $y_1 \leq y_2 \leq \cdots$. Furthermore, let $S^* = \{s_1^*, s_2^*, \ldots, s_K^*\}$ be any finite solution (ordered), i.e., with $s_1^* < \cdots < s_K^*$. It is sufficient to show that $|Y^*| \leq K$, since from part 3 of Lemma 5.6 we then know that Y^* is feasible, and better than the finite solution S^*. To this end, we first prove that

$$\text{cover}\{s_k^* : k < i\} \cap D_{total} \subseteq \text{cover}\{y_k : k < i\} \cap D_{total}.$$

$i = 1$: $\emptyset = \emptyset$.
$i \to i+1$: From the induction hypothesis we know that

$$\text{cover}\{s_k^* : k < i\} \cap D_{total} \subseteq \text{cover}\{y_k : k < i\} \cap D_{total},$$

yielding $D_i \subseteq D_{total} \setminus \text{cover}\{s_k^* : k < i\}$.

Claim 1: $y_i \geq s_i^*$.

Suppose $s_i^* > y_i$. Since $y_i = L(D_i) = \inf\{l_d : d \in D_i\}$ and $s_i^* > y_i$ there exists some $d \in D_i$ with $l_d < s_i^*$. According to (5.2) this means d is **not** covered by s_k^* if $k \geq i$. Since $d \in D_i$ implies that

$$d \notin \text{cover}(\{s_k^* : k < i\})$$

it is also **not** covered by s_k^* if $k < i$, a contradiction.

Claim 2: $\text{cover}(s_i^*) \cap D_i \subseteq \text{cover}(y_i)$.

Let $x \in \text{cover}(s_i^*) \cap D_i$. Since $x \in \text{cover}(s_i^*)$ we get $f_d \leq s_i^*$ and using Claim 1, $f_d \leq y_i$. On the other hand, $d \in D_i$ and the definition of $y_i = L(D_i)$ yields $y_i \leq l_d$. Together, $f_d \leq y_i \leq l_d$ such that $y_i \in \mathcal{T}(d)$ meaning that $d \in \text{cover}(y_i)$ (see (5.2)).

The induction hypothesis together with Claim 2 shows the result.

Finally, since S^* is a finite solution we get that $D_{total} \subseteq \text{cover}\{s_i^* : i \leq K\}$ and hence $D_{total} \subseteq \text{cover}\{y_i : i \leq K'\}$, where $K' = \min\{|Y^*|, K\} \leq K$. This means that at the end of iteration K' the set

$$\begin{aligned} D = D_{K'} &= D_{K'-1} \setminus \text{cover}\{y_{K'}\} \\ &= D_{total} \setminus \text{cover}\{y_1, \ldots, y_{K'}\} = \emptyset \end{aligned}$$

and the algorithm terminates with a solution with cardinality $|Y^*| = K' \leq K$. □

The algorithm can be transfered to arbitrary graphs, if we know in advance, that no point $d \in D_{total}$ can be covered by two stations belonging to different edges of G. This is formalized below.

Notation 5.8. *For* $e_1, e_2 \in E$ *let* $\text{conf}(e_1, e_2) = \text{cover}(e_1) \cap \text{cover}(e_2)$ *be the* **conflict zone** *of edge* e_1 *and edge* e_2.

Then we get the following result.

Theorem 5.9. *If* $\text{conf}(e_1, e_2) = \emptyset$ *for all* $e_1, e_2 \in E$, *an optimal solution of (CSL-region) is given by*

$$S^* = \bigcup_{e \in E} S_e^*,$$

where S_e^* *is the output of Algorithm 11 for edge* e.

Proof. From the assumption we know that each point $d \in D_{total}$ can be covered from at most one edge $e \in E$. This means that the assumptions of Corollary 3.29 hold. Realizing that Theorem 3.28 and Corollary 3.29 hold also for a non-finite set D_{total} instead of the finite set $\mathcal{D}$ used in the proof, we obtain the result together with Theorem 5.7. □

Fortunately, in our real-world data the total amount of population living in a conflict zone is rather small. For the Euclidean unweighted stop location problem we calculated the following data:

- The union of circles with a radius of 2 km centered at the existing stops covers 52.4 % of the total population.
- We computed that cover($\mathcal{T}$) contains 65.20% of the population, i.e., only 65.20 % of the inhabitants of Germany live within a distance of less than 2 km from the *Deutsche Bahn* railway network (nodes and edges). Thus f_{cover} can be increased by not more than 12.80 %.
- The conflict zones are certainly not empty in our practical problem instance, but they only contain 1.5 % of the total population.

If we discard the population in the conflict zones from the covering problem we can apply Algorithm 11. With this approach we at least obtain a lower bound on (CSL-region), since the generated solution will probably not cover all of the discarded inhabitants. On the other hand, we can still achieve 80% of the possible improvement of f_{cover} such that we conclude that Algorithm 11 is applicable for practical problem instances.

5.2 Minimizing the Total Door-to-door Travel Time

We now turn our attention back to demand points, but focus on a completely different objective function, namely the total *door-to-door travel time* over all customers. The door-to-door travel time includes not only the *travel time* while sitting in a bus or train, but also the *access time* to reach the first station and to get from the last station of the trip to the final destination. Instead of determining the total door-to-door travel time, we only calculate its total change. The changes in the door-to-door travel time arise due to a positive and a negative effect. The positive effect takes into account that new stops decrease the distance to the stations for some of the customers. These customers hence have a smaller access time to reach their first station. For the sake of simplicity we assume in this model that all customers depart from their closest station. On the other hand, the negative effect is the time which arises by the additional stopping activities of the trains. This value has been considered in the models before, and was denoted by f_{time}.

We now describe how the reduction of the access time f_{access} can be calculated. To this end, let S^{ex} be the already existing stations, and let S be a possible set of new stops.

For a demand point $d \in \mathcal{D}$,

- $\gamma_d(d, S^{ex})$ is the closest distance to one of the existing stations, i.e., without opening any new stop.
- $\gamma_d(d, S^{ex} \cup S)$ is the closest distance to a station after the new stops have been opened.
- The reduction of the distance for customers of demand point d hence is $\gamma_d(d, S^{ex}) - \gamma_d(d, S^{ex} \cup S)$, which may be zero (if no new stop is closer to d than the closest existing stop) or positive.

To transform a possible reduction of the distance into an amount of saved access time we introduce a piecewise linear function in two variables $g : \mathbb{R} \times \mathbb{R} \to \mathbb{R}$, assigning an amount of saved time to each reduction of the distance, given as a pair consisting of the old and the new distance of a demand point to its closest train stop. For g we require that

$$x \geq y \text{ implies } g(x, y) \geq 0.$$

For example, g can be defined as

$$g(x, y) = (x - y)/5,$$

assuming an average (walking) speed of 5 km/h, or as

$$g(x,y) = \begin{cases} \frac{x-y}{4} & \text{if } x \leq 1 \text{ (the customer walks)} \\ \frac{x-y}{7} & \text{if } 1 < x \leq 5 \text{ (the customer uses a bike)} \\ \frac{x-y}{20} & \text{if } 5 < x \text{ (the customer uses a bus or a car).} \end{cases}$$

Note that this definition assumes that a customer stays with the same means of transport used for the old distance x.

The positive effect of new stations S on the door-to-door travel time through saved access time can hence be calculated by

$$f_{\text{access}}(S) = \sum_{d \in \mathcal{D}} w_d g(\gamma_d(d, S^{ex}), \gamma_d(d, S^{ex} \cup S)).$$

The door-to-door travel time model can now be formulated.

$$\min f_{\text{time}}(S) - f_{\text{access}}(S)$$

such that

$$S \subseteq \mathcal{T}.$$

Note that in this model we neglect the change in train riding time that is caused by starting or ending the trip at a different train stop, assuming that these gains and losses roughly even out. We can now define the corresponding *door-to-door travel time stop location problem.*

(DSL)

Given $G = (V, E)$ with its set of points $\mathcal{T}$ and with traffic loads c_e for all $e \in E$, c_v for all $v \in V$, as well as a finite set of points $\mathcal{D}$ with weights w_d and gauges γ_d for all $d \in \mathcal{D}$, find a set $S \subseteq \mathcal{T}$ such that

$$f_{\text{DSL}}(S) = f_{\text{time}}(S) - f_{\text{access}}(S)$$

is minimized.

The door-to-door travel time stop location problem is the first continuous stop location problem that has been mentioned in the literature, see [HLS+01]. Some important results have already been obtained in this paper, including the NP-hardness of (DSL) which we show next.

Theorem 5.10. *(DSL) is NP-hard.*

Proof. We reduce the unweighted (CSL), which is NP-hard according to Theorem 3.2, to (DSL). Given an instance of the unweighted (CSL) with $\mathcal{D}$, $\mathcal{T}$, γ_d and $K \leq |\mathcal{D}|$, let $M > |\mathcal{D}|$ and define

$$\begin{aligned} S^{ex} &= \emptyset, \\ w_p &= M \text{ for all } d \in \mathcal{D}, \quad \text{and} \\ c_e &= 1 \text{ for all } e \in E. \end{aligned}$$

Furthermore, define

$$g(x,y) = \begin{cases} 1 & \text{if } y \leq r \\ 0 & \text{otherwise.} \end{cases}$$

With these definitions, we get that

$$f_{\mathrm{DSL}}(S) = f_{\mathrm{time}}(S) - f_{\mathrm{access}}(S) = |S| - M|\mathrm{cover}(S)|.$$

Claim: There exists a solution S to (DSL) with $f_{\mathrm{DSL}}(S) \leq K - M|\mathcal{D}|$ if and only if there exists a solution to (CSL) with no more than K stops.

$\Longrightarrow$: First, let S be a solution of (DSL) with $f_{\mathrm{DSL}}(S) \leq K - M|\mathcal{D}|$. We show that then $\mathrm{cover}(S) = \mathcal{D}$: Assume to the contrary that

$$|\mathrm{cover}(S)| \leq |\mathcal{D}| - 1. \tag{5.3}$$

Then

$$\begin{aligned} f_{\mathrm{DSL}}(S) &= |S| - M|\mathrm{cover}(S)| \\ &\geq -M|\mathrm{cover}(S)| \\ &\geq M - M|\mathcal{D}| \quad \text{due to (5.3)} \\ &> |\mathcal{D}| - M|\mathcal{D}| \\ &\geq K - M|\mathcal{D}| \\ &\geq f_{\mathrm{DSL}}(S), \end{aligned}$$

a contradiction. Furthermore, $\mathrm{cover}(S) = \mathcal{D}$ implies

$$\begin{aligned} |S| - M\mathrm{cover}(S) &= f_{\mathrm{DSL}}(S) \\ &\leq K - M|\mathcal{D}| \\ &= K - M|\mathrm{cover}(S)| \end{aligned}$$

and hence $|S| \leq K$.

$\Longleftarrow$: For the other direction, let S be a solution of (CSL) with $\text{cover}(S) = \mathcal{D}$ and $|S| \leq K$. Hence,

$$\begin{aligned} f_{\text{DSL}}(S) &= |S| - M|\text{cover}(S)| \\ &= |S| - M|\mathcal{D}| \\ &\leq K - M|\mathcal{D}| \end{aligned}$$

which completes the proof. □

Note that we obtain the following result on the objective functions.

Lemma 5.11. *Let $S_1, S_2 \subseteq \mathcal{T}$ be two disjoint sets of stops, i.e., $S_1 \cap S_2 = \emptyset$. Then*

$$\begin{aligned} f_{\text{time}}(S_1 \cup S_2) &= f_{\text{time}}(S_1) + f_{\text{time}}(S_2) \\ f_{\text{access}}(S_1 \cup S_2) &\leq f_{\text{access}}(S_1) + f_{\text{access}}(S_2) \\ f_{\text{DSL}}(S_1 \cup S_2) &\geq f_{\text{DSL}}(S_1) + f_{\text{DSL}}(S_2). \end{aligned}$$

Proof.

1.

$$\begin{aligned} f_{\text{time}}(S_1 \cup S_2) &= \sum_{s \in S_1} c_{g(s)} + \sum_{s \in S_2} c_{g(s)} \\ &= f_{\text{time}}(S_1) + f_{\text{time}}(S_2). \end{aligned}$$

2. For f_{access} let

$$f^d_{\text{access}}(S) = w_d g(\gamma_d(d, S^{ex}), \gamma_d(d, S^{ex} \cup S))$$

denote the positive effect in the access time for a demand point $d \in \mathcal{D}$. For $d \in \mathcal{D}$ we distinguish the following cases:

Case 1: $\gamma_d(d, S^{ex} \cup S_1 \cup S_2) = \gamma_d(d, S^{ex})$, i.e., the closest stop to d does not change by adding new stops. Then

$$\begin{aligned} g(\gamma_d(d, S^{ex}), \gamma_d(d, S^{ex} \cup S_1 \cup S_2)) &= g(\gamma_d(d, S^{ex}), \gamma_d(d, S^{ex} \cup S_1)) \\ &= g(\gamma_d(d, S^{ex}), \gamma_d(d, S^{ex} \cup S_2)) \end{aligned}$$

and hence $f^d_{\text{access}}(S_1 \cup S_2) \leq f^d_{\text{access}}(S_1) + f^d_{\text{access}}(S_2)$.

Case 2: $\gamma_d(d, S^{ex} \cup S_1 \cup S_2) = \gamma_d(d, S_1)$, i.e., there is a new stop (in S_1) which is closer to d than all existing stops. Then

$$g(\gamma_d(d, S^{ex}), \gamma_d(d, S^{ex} \cup S_1 \cup S_2)) = g(\gamma_d(d, S^{ex}), \gamma_d(d, S^{ex} \cup S_1)),$$

and since

$$g(\gamma_d(d, S^{ex}), \gamma_d(d, S^{ex} \cup S_2)) \geq 0$$

(due to $\gamma_d(d, S^{ex}) \geq \gamma_d(d, S^{ex} \cup S_2)$) we obtain

$$f^d_{\text{access}}(S_1 \cup S_2) \leq f^d_{\text{access}}(S_1) + f^d_{\text{access}}(S_2).$$

Together we get

$$\begin{aligned} f_{\text{access}}(S_1 \cup S_2) &= \sum_{d \in \mathcal{D}} f^d_{\text{access}}(S_1 \cup S_2) \\ &\leq \sum_{d \in \mathcal{D}} \left(f^d_{\text{access}}(S_1) + f^d_{\text{access}}(S_2)\right) = f_{\text{access}}(S_1) + f_{\text{access}}(S_2). \end{aligned}$$

3. Finally,

$$\begin{aligned} f_{\text{DSL}}(S_1 \cup S_2) &= f_{\text{time}}(S_1 \cup S_2) - f_{\text{access}}(S_1 \cup S_2) \\ &\geq f_{\text{time}}(S_1) + f_{\text{time}}(S_2) - f_{\text{access}}(S_1) - f_{\text{access}}(S_2) \\ &= f_{\text{DSL}}(S_1) + f_{\text{DSL}}(S_2). \end{aligned}$$

□

In our experimental study, described in [HLS+01] we used a genetic algorithm (see, e.g., [Gol89]) to find a heuristic solution for (DSL). This was done by choosing a set of candidates $\mathcal{S}^{cand} \subseteq \mathcal{T}$. Each feasible solution $S \subseteq \mathcal{S}^{cand}$ of (DSL) is then described by a vector $x \in \{0,1\}^{|\mathcal{S}^{cand}|}$ by

$$x_s = \begin{cases} 1 & \text{if } s \text{ is contained in } S \\ 0 & \text{otherwise.} \end{cases}$$

$\mathcal{S}^{cand}$ can be chosen by equally distributing possible candidates along all edges $e \in E$, or by using one of the finite dominating sets $\mathcal{S}$ or $\mathcal{S}^*$ developed in Sections 3.2 and 4.4. Before the genetic algorithm is started, the set of candidates is reduced according to the following observation.

Lemma 5.12. *Let $s^0 \in \mathcal{S}^{cand}$. If*

$$f_{\text{time}}(\{s^0\}) > f_{\text{access}}(\{s^0\}),$$

then no optimal solution to (DSL) will contain s^0.

Proof. Let $S \subseteq \mathcal{S}^{cand}$ be a feasible solution of (DSL) and let $s^0 \in S$. Define $S' = S \setminus \{s^0\}$. Then, due to Lemma 5.11,

$$\begin{aligned} f_{\text{DSL}}(S) &\geq f_{\text{DSL}}(S') + f_{\text{DSL}}(\{s^0\}) \\ &= f_{\text{DSL}}(S') + f_{\text{time}}(\{s^0\}) - f_{\text{access}}(\{s^0\}) \\ &> f_{\text{DSL}}(S'), \end{aligned}$$

i.e., a solution containing s^0 can be strictly improved by subtracting s^0 and hence s^0 will never appear in an optimal solution. □

The starting population of the genetic algorithm was chosen randomly, and cross-over and mutation was performed according to the usual rules. The set of candidates used for the genetic algorithm consisted of 6 700 potential new stops after the reduction according to Lemma 5.12. Assuming that two minutes is the time needed for an additional stop and using the function $g(x, y) = (x - y)/5$ to describe the gain in access time when the distance to the nearest train stop changes from x to y, we started the genetic algorithm with this set of candidates and with three different starting solutions, three times each. These starting solution were chosen randomly, but with three different probabilities to establish a candidate as a stop, namely with $p \in \{0.25, 0.5, 0.75\}$, resulting in three initial populations, P_1,P_2, and P_3, where

- P_1 contained solutions with an average of 1 700 new stops,
- P_2 contained solutions with an average of 3 350 new stops, and
- P_3 contained solutions with an average of more than 5 000 new stops.

We used a population size of 20, and the probability p_m for mutation of bits after crossover was set to 0.0001. After 100 generations each, we let the resulting population with the best value for f_{DSL} evolve for another 900 generations. The development of this population over the course of its 1 000 generations is shown in Figure 5.4.

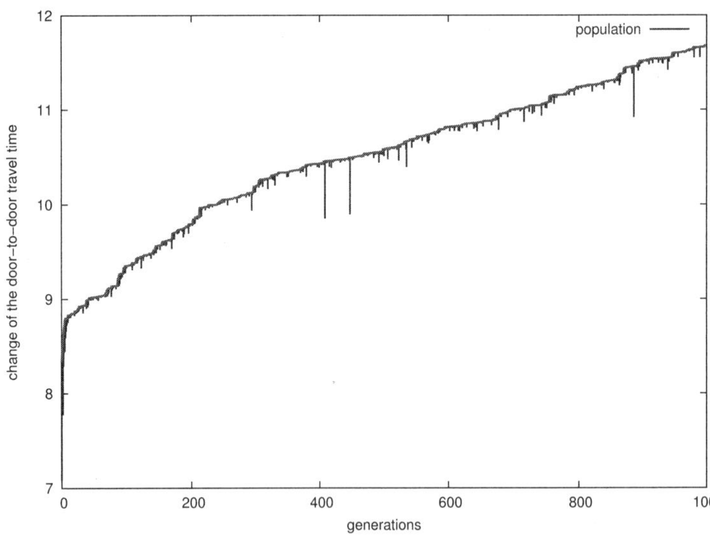

Fig. 5.4. Development of the population in the genetic algorithm for the door-to-door travel time.

It turned out that the genetic algorithm converges to a stable number of new stops very quickly, i.e., we actually do not need to first determine a good probability p for the creation of the starting population: While a starting population with varying initial probabilities p_b contained individuals with almost no candidates as well as individuals with almost all of the 6 700 candidates, the difference between the lowest and the highest number of candidates in the individuals of the 10th generation had already shrunk to less than 1 000, and after 100 generations this difference was merely 22.

Other Objective Functions

Finally, from the point of view of a public transportation company it is desirable to consider demand changes and their impact on the income of the company as follows. Assume that for customers traveling from d_1 to d_2 the door-to-door travel time increased. Then take an elasticity factor estimating how many of these customers will change to another means of transport and estimate the loss in income obtained by this change. On the other hand, if the door-to-door travel time decreases for possible trips, we have to estimate how many other customers will be attracted by the better connection and estimate the money they will spend. For this model a lot of other data has to be taken into account:

- a matrix containing the number of (potential) customers for all possible pairs of demand points $d_1, d_2 \in \mathcal{D}$. This data is needed to calculate the changes in the door-to-door travel time for each possible trip. (Note that for (DSL) it is enough to know the *sum* of the changes in the door-to-door travel time, which can be estimated without detailed information.)
- the routes customers are going to use (for calculating the additional time of the stopping activities) and also the changes they make if their closest station changes
- elasticity factors for demand changes
- ticket prices
- additional costs arising because of the longer running times of the trains (for drivers, other crew members, or for the rolling stock itself)
- fixed costs of the new stops.

Since an extension along these lines gets more and more complicated, a further development of metaheuristic approaches like the genetic algorithm described in this chapter seems to be appropriate in this context.

Part II

Delay Management

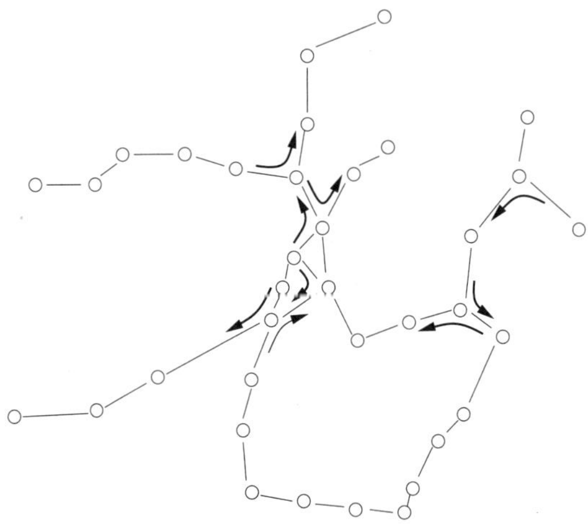

6

Introduction

A major reason for complaints about public transportation is the missing punctuality, which — unfortunately — is a fact in many transportation systems. Since it seems to be impossible to avoid delays completely, it is a necessary issue in the dispositive work of a public transportation company to deal with delayed vehicles.

We focus on the convenience of the customers, so let us first analyze the effects of a delayed vehicle on its passengers. If a vehicle reaches a station with a delay, one consequence is that customers getting out there will reach their destination with this delay. This is annoying, but it is not worth a complaint if the delay is rather small. The situation becomes worse if customers who wish to change from the delayed vehicle into another bus or train miss their connection, and this can happen even in the case of small delays.

Let us now consider some vehicle (e.g., a train g) that arrives at a station with a delay. At the station, there are other vehicles (e.g., two buses h and h') ready to depart, see Figure 6.1. What should each of these connecting vehicles do? There are two alternatives:

- A connecting bus can **wait** and therefore cause delay for the customers within the bus, but also for the customers who wish to get on this bus later on, and possibly for subsequent other buses which will have to wait for its delay.
- On the other hand, if a connecting bus **departs** on time, all customers who planned to change from the delayed train into the bus will miss their connection.

In the first case the connecting vehicle does not depart at its scheduled time, but with a delay. The new departure time of this vehicle is called its *perturbed* timetable. In the second case, the *perturbed* departure time of the bus equals the scheduled one.

The *delay management problem* is to find wait-depart decisions and a perturbed timetable in case of some known delays, not only for one single bus,

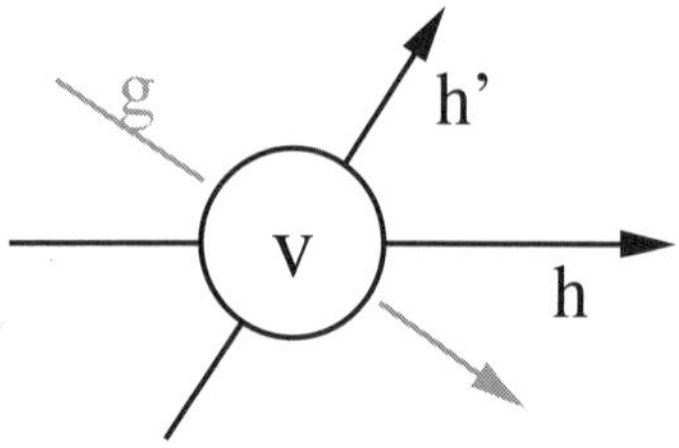

Fig. 6.1. The wait-depart decision at one single station.

but for all vehicles in the public transportation network, such that the "inconvenience" over all customers is minimized. As inconvenience we consider the delay of the customers and the number of missed connections.

To avoid that customers miss their connections, one could force all departing vehicles to wait until all delayed vehicles have arrived. This makes sure that all connections are maintained, but leaves the transportation system with many delayed vehicles. Consequently, the delay spreads out through the whole network, and hence will affect many customers. On the other hand, if all vehicles would depart as early as possible the number of delayed vehicles is minimized, but in this case many customers miss their connections.

A solution between these two extremes can be obtained by using the following models.

- In the *delay management problem with fixed connections (TT)* we assume that we already know the wait-depart decisions and we try to find a perturbed timetable which is as close as possible to the original one. This problem can be solved easily, but it will turn out that it is an important building block in the following models.
- In the *total delay management problem (TDM)* we consider the inconvenience of a customer as the amount of delay when he arrives at his destination, and minimize the sum of all delays over all customers within the system. Since the delay in case of a missed connection is usually large, but many delayed vehicles affect many customers, the solution found will allow some vehicles to depart on time and force other vehicles to wait.
- The *bicriteria delay management problem (BDM)* investigates the two objective functions
  - minimize the delay of all vehicles, and
  - minimize the number of missed connections

  simultaneously, trying to find efficient solutions with respect to both criteria.
- It will turn out that all these models are special cases of the *general delay management problem (GDM)*, which is also defined by two criteria. In this problem we want to minimize

- the number of customers missing a connection and
- the amount of the additional delay of the remaining customers.

Chapter 6 is structured as follows: In this chapter we mention our application and then give an overview of related literature. Furthermore, we introduce the notation and definitions needed to state the delay management problem formally. Especially we discuss perturbed timetables and give two integer programming descriptions of the set of all feasible perturbed timetables. The first one is based on the "intuitive" description of the problem, while the second one uses the concept of event-activity networks. A detailed description of this concept is given.

6.1 Application

The subject of delay management was brought up by two large traffic associations serving the states Rheinland-Pfalz and the Saarland (both in Germany) within a project supported by *Stiftung Rheinland-Pfalz für Innovation*, see [SS01]. Public transportation companies are interested in analyzing the consequences of delays, or, more generally, also of (small) changes in the schedule. When analyzing such changes, it is important not to look only at one single transportation company, but to take into account also connections between different public transportation companies. The following two problems are of special interest in our applications.

On-line wait-depart decisions: On a regional train line in Rheinland-Pfalz, the 40 km long *Lautertalbahn* leading from Kaiserslautern to Lauterecken, *Deutsche Bahn* installed an automatic system informing the bus drivers waiting at the stations about the exact arrival times of the incoming trains. The question arising now for the drivers is, whether they should wait for a delayed train or depart on time. This decision will be investigated from the perspective of the customers.

Short-term adaption of timetables: Suppose there is a large construction area somewhere in a city, leading to a detour for some bus lines. Then it is often known that each time a bus goes there it will gain a delay of, say, five minutes. Since this delay will occur as long as the construction area is present it is worthwhile to design a new *perturbed* timetable for this period of time. It is important that in this new timetable no bus departs earlier than planned, since this would annoy passengers who maybe have not seen any announcement about the perturbed timetable. How to calculate such a timetable in a way that many customers are satisfied will also be discussed.

It will turn out that we can use the same mathematical model for analyzing both problems. In our test data we analyze the effects of delays of trains of the Lautertalbahn on subsequent other trains and buses belonging to different public transportation companies. The test set consists of

- 823 stations,
- 1314 vehicles, and
- 2118 direct rides.

We also have to specify the set of connections we consider. We use four different sets $\mathcal{U}_5$, $\mathcal{U}_{10}$, $\mathcal{U}_{30}$, and $\mathcal{U}_{60}$, where set $\mathcal{U}_x$ contains reasonable connections with a scheduled waiting time of less than x minutes. By "reasonable" we mean that we do not consider connections where changing results in going directly back to the previous station. The sizes of the sets U_x are:

$$\begin{aligned}|\mathcal{U}_5| &= 6531\\ |\mathcal{U}_{10}| &= 10659\\ |\mathcal{U}_{30}| &= 39371, \text{ and}\\ |\mathcal{U}_{60}| &= 80716.\end{aligned}$$

The resulting event-activity network (which will be introduced in Section 6.4) has a size of 46720 nodes (events). The number of edges (activities) depends on the set U_x used and varies between 51937 (for U_5) and and 126122 (for U_{60}). This data set is the basis for the numerical results mentioned in this part.

6.2 Related Literature

Since in the delay management problem new departure times for each vehicle at each station have to be determined, it is closely related to the problem of finding timetables in public transportation. In this field, a lot of research has been done. Timetabling is fairly easy if the timetables need not be periodic, but even the detection of a feasible periodic timetable is an NP-hard problem (see [SU89, Nac98]). Many approaches were applied in this area. Integer programming approaches and their combinatorial analysis were studied in [BLNN98, Nac98, Nac97, Odi96, Kri96, Nac94, Ste88]. Extensions of these approaches using cycle bases were obtained in [LR05, Lie03, PK03, Pee02, PK01]. Furthermore, quadratic semi-assignment models were discussed in [AC97, Dom89, Fle91, DV95, KS87]. For graph-theoretical approaches we refer to [Wei81, Kri96, Nac96]. Moreover there exist geometric approaches (polygon-on-circle model) in [Bur86, BBH90, BH86], network design models in [BD92, BT81, Voß92] and various heuristics and metaheuristics, see [LPW04, XC94, Car98, Car99, KNV96, NV96, NV97]. Other models which were also considered in scheduling and timetabling are the max-plus-algebra (e.g., used by [GBO99, Gov98b, Gov98a]) or the heaps-of-pieces-approach, see [vE01]. Practical experience is — among others — reported in

[Kri96, NV96, DV95, Ste88, HH87, Gün85]. In [LM02], a case study (of Berlin underground) is presented.
As objective functions, the total waiting time or the travel time of the passengers have mainly been considered, sometimes also the costs of operating the timetable. The main difference between finding a timetable and solving the delay management problem, besides the slightly different objective functions, is that in the delay management problem the connections are not given in advance. In fact, the main decision that must be made in delay management is to decide which connections should be maintained and which can be dropped.

Delays and their consequences have been addressed before in a few papers, see e.g., [HK81, HK98b] for models to calculate expected delays. The effect of delays on the robustness of the timetables was discussed, e.g., in [EFK01a, EFK01b, EF02]. Decreasing the expected delay by investing in new tracks was investigated in [EFK01a]. Analyzing delays using the max-plus-algebra was done in [HdV01]. Max-plus-algebra together with stochastic processes and the theory of option pricing have recently been used to model the delay management problem, see [Bau05].

How to *react* in case of delays has, due to the size and complexity of the problem, been mainly tackled by simulation and expert systems. We refer to [SM97, SM99, SBK01, SMBG01] for providing a knowledge-based expert system including a simulation of wait-depart decisions with a *what-if* analysis. Simulation including the capacity constraints on the tracks has also been used in [Ack99, SM01].

Only few optimization models for the delay management problem have been published. In [ADGGT99] an objective function from the customers' point of view is mentioned. Such an objective was taken into account in the models developed independently by [Kli00a, SBK01] and by [Sch01b]. Both studied an approximation of the effects of delays on the customers and presented non-linear mixed-integer programming formulations. It will turn out that their models are equivalent to a special case of the total delay management problem which will be formulated as a linear program in Section 8.4 of this text. In [APW02], the wait-depart decision at one single station with an unknown amount of delay was analyzed.

Our basic delay management models (TDM) will be developed in Chapter 8. (Note that a short review about the corresponding linear model in Section 8.1 was published in [Sch01c].) These models already had some influence on the research community, which is reflected in several papers dealing with further results and extensions. An important contribution is given in [GJPS05], where the complexity status of the delay management problem (TDM) as it will be defined in Chapter 8 has been clarified. It turns out that even special cases are NP-hard. For more complexity issues we also refer to [GGJ$^+$04]. An on-line version of the delay management problem has been investigated in [GJPW06].

Recently, [GHL06] developed some interesting reformulations and extensions of our models.

6.3 A Model for the Delay Management Problem

In this section we specify the notation necessary for dealing with the delay management problem. Let PTN $= (V, E)$ and F be the set of vehicles. For each vehicle g, we introduce

- $V^g \subseteq V$ as the set of stations where it stops, and
- $E^g \subseteq V^g \times V^g$ as the set of its direct rides within the transportation network.

If it is possible to change from vehicle g to vehicle h at a station v, then (g, h, v) will be called a *connection* (see Figure 6.2) and the whole set of connections is given by

$$\mathcal{U} \subseteq F \times F \times V.$$

We can assume that each vehicle $g \in F$ stops at the same station only once, otherwise we use a time-dependent representation of the respective station.

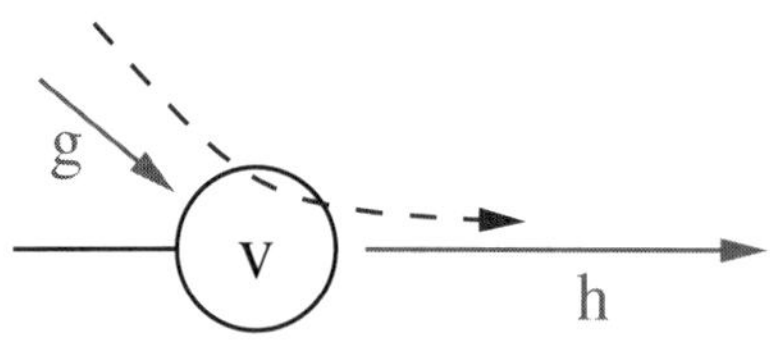

Fig. 6.2. A connection (g, h, v) from vehicle g to vehicle h at station v.

Definition 6.1. *A* **timetable** Π *is given by natural numbers* $\Pi_{arr}{}_g^v, \Pi_{dep}{}_g^v$ *for all* $g \in F, v \in V^g$ *with the following meaning:*

- $\Pi_{arr}{}_g^v$ *is the scheduled arrival time of vehicle* g *at station* v*, and*
- $\Pi_{dep}{}_g^v$ *is the scheduled departure time of vehicle* g *at station* v*.*

Furthermore, a timetable Π *is* **feasible**, *if it satisfies conditions (6.1),(6.3), and (6.5) below.*

Waiting requirements: Consider vehicle g at station v.

- Let $L_g^v \in \mathbb{N}$ be the given minimal waiting time of vehicle g, that allows passengers to get off and on vehicle g at station v, or makes sure that the driver gets a break.

- The scheduled waiting time of vehicle g at station v is $\Pi_{dep}{}_g^v - \Pi_{arr}{}_g^v$. In a feasible timetable Π we require that

$$\Pi_{dep}{}_g^v - \Pi_{arr}{}_g^v \geq L_g^v. \tag{6.1}$$

- The slack time is denoted by s_g^v and is given by

$$s_g^v = \Pi_{dep}{}_g^v - \Pi_{arr}{}_g^v - L_g^v. \tag{6.2}$$

Driving requirements: Consider vehicle g going from station v to station u.

- Let $L_g^{vu} \in \mathbb{N}$ be the minimal driving time when vehicle g drives as fast as possible between stations v and u.
- The scheduled driving time is $\Pi_{arr}{}_g^u - \Pi_{dep}{}_g^v$. In a feasible timetable, we require

$$\Pi_{arr}{}_g^u - \Pi_{dep}{}_g^v \geq L_g^{vu}. \tag{6.3}$$

- The slack time on this driving edge is denoted by s_g^{vu} and is given by

$$s_g^{vu} = \Pi_{arr}{}_g^u - \Pi_{dep}{}_g^v - L_g^{vu}. \tag{6.4}$$

Changing requirements: Finally, consider a customer transferring from vehicle g to vehicle h at station v, i.e., a connection $(g, h, v) \in \mathcal{U}$.

- The minimal necessary changing time for getting off vehicle g, walking to the departure place of vehicle h, and boarding is denoted by $L_{gh}^v \in \mathbb{N}$.
- The scheduled time allowed for changing is $\Pi_{dep}{}_h^v - \Pi_{arr}{}_g^v$, and again in a feasible timetable it is required that

$$\Pi_{dep}{}_h^v - \Pi_{arr}{}_g^v \geq L_{gh}^v. \tag{6.5}$$

- The slack time for changing is denoted by s_{gh}^v and is given by

$$s_{gh}^v = \Pi_{dep}{}_h^v - \Pi_{arr}{}_g^v - L_{gh}^v. \tag{6.6}$$

Since the given timetable and the minimal necessary times for waiting, driving, and changing are integer vectors, it holds that also all slack times are integer.

By $d_g^v \in \mathbb{N}$ let us denote the *source delay* (or *primary delay*) of vehicle g when arriving at station v. (In practice, this delay is usually given in minutes.) I.e.,

$$\mathcal{E}_{del} = \{(g, v) : d_g^v > 0\}$$

is the set of all delayed arrivals known as input for the optimization.

If a delay occurs, say vehicle g arrives at some station v with a source delay of $d_g^v > 0$, then simply increasing $\Pi_{arr}{}_g^v$ by d_g^v will in general make the timetable infeasible. The goal now is to find a *wait-depart decision* for all connections and a *feasible perturbed timetable* $x_{arr}{}_g^v, x_{dep}{}_g^v$ for all $g \in F, v \in V^g$ which is best with respect to the objective function under consideration. Note that a perturbed timetable is feasible, if the waiting requirements (6.1) and the driving requirements (6.3) are satisfied, i.e., the delay of a vehicle is carried over to its next station correctly. Since we do not force departing vehicles to wait for possibly delayed arriving vehicles, we do not require that a feasible perturbed timetable also satisfies the changing requirements (6.5). Formally, we define a perturbed feasible timetable as follows.

Definition 6.2. *Given a timetable Π, a set of source delays $d_g^v \geq 0$, a* **perturbed timetable** $x_{arr}{}_g^v, x_{dep}{}_g^v \in \mathbb{N}$ *for all* $g \in F, v \in V^g$ *is* **feasible**, *if the following conditions hold.*

$$x_{arr}{}_g^v \geq \Pi_{arr}{}_g^v + d_g^v \quad \text{for all } g \in F, v \in V^g, \tag{6.7}$$

$$x_{dep}{}_g^v \geq \Pi_{dep}{}_g^v \quad \text{for all } g \in F, v \in V^g, \tag{6.8}$$

$$x_{dep}{}_g^v - x_{arr}{}_g^v \geq L_g^v \quad \text{for all } g \in F, v \in V^g, \tag{6.9}$$

$$x_{arr}{}_g^u - x_{dep}{}_g^v \geq L_g^{vu} \quad \text{for all } g \in F, (v,u) \in E^g. \tag{6.10}$$

Conditions (6.7) and (6.8) make sure that the perturbed timetable respects the source delays, while in (6.9) we require that a vehicle arriving at a station with a delay also departs with this delay, and (6.10) ensures that a vehicle departing at some station with a delay carries over that delay to its next station. In both cases the delay may be reduced at most by the given slack time. The situation of conditions (6.9) and (6.10) is depicted in Figure 6.3 a) and b), respectively. Note that in a perturbed timetable we do **not** require that all connections are maintained, i.e., (6.5) need not be satisfied.

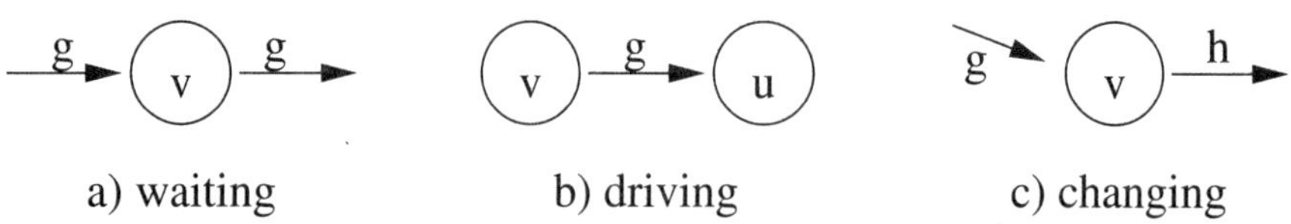

Fig. 6.3. Illustration of conditions (6.9),(6.10), and of constraint (6.15).

Finally, we make precise the wait-depart decisions.

Definition 6.3. *A connection $(g, h, v) \in \mathcal{U}$ at station v is called* **maintained** *if vehicle h waits for vehicle g, i.e., if*

$$x_{dep}{}_h^v - x_{arr}{}_g^v \geq L_{gh}^v.$$

Otherwise, the connection is called **missed**.

For describing the set of feasible solutions we have to specify the relation between the wait-depart decisions and the perturbed timetable. To this end we use variables with the following meaning:

$$y_{dep}{}_g^v = \text{departure delay of vehicle } g \text{ at station } v$$
$$y_{arr}{}_g^v = \text{arrival delay of vehicle } g \text{ at station } v$$
$$\bar{z}_{ghv} = \begin{cases} 0 & \text{if connection } (g,h,v) \in \mathcal{U} \text{ is maintained} \\ 1 & \text{if connection } (g,h,v) \in \mathcal{U} \text{ is missed.} \end{cases}$$

It will be more convenient to use only the additional delay y instead of the perturbed timetable x, which is given by

$$x_{dep}{}_g^v = \Pi_{dep}{}_g^v + y_{dep}{}_g^v$$
$$x_{arr}{}_g^v = \Pi_{arr}{}_g^v + y_{arr}{}_g^v$$

for all $g \in F, v \in V^g$.
For the additional delay y, Definition 6.3 then can be rewritten as follows: A connection $(g,h,v) \in \mathcal{U}$ is called **maintained** if

$$y_{arr}{}_g^v - y_{dep}{}_h^v \leq s_{gh}^v. \tag{6.11}$$

Let M be a sufficiently large parameter. (Note that $M = \max_{g \in F, v \in V} d_g^v$ is large enough, as we will show later on.) Then $(y_{dep}, y_{arr}, \bar{z})$ is feasible, if the following constraints are satisfied.

$$y_{arr}{}_g^v \geq d_g^v \quad \forall (g,v) \in \mathcal{E}_{del} \tag{6.12}$$
$$y_{arr}{}_g^v - y_{dep}{}_g^v \leq s_g^v \quad \text{for all } g \in F, v \in V^g \tag{6.13}$$
$$y_{dep}{}_g^v - y_{arr}{}_g^u \leq s_g^{vu} \quad \text{for all } v,u \in V^g, g \in F : (v,u) \in E^g \tag{6.14}$$
$$-M\bar{z}_{ghv} + y_{arr}{}_g^v - y_{dep}{}_h^v \leq s_{gh}^v \quad \text{for all } (g,h,v) \in \mathcal{U} \tag{6.15}$$
$$y_{dep}{}_g^v, y_{arr}{}_g^v \in \mathbb{N} \quad \text{for all } g \in F, v \in V^g$$
$$\bar{z}_{ghv} \in \{0,1\} \quad \forall (g,h,v) \in \mathcal{U}.$$

Constraints (6.12) establish the source delays which we assume to be known for all arrivals and which are strictly positive for all $(g,v) \in \mathcal{E}_{del}$. The delay management problem is only interesting for at least one source delay $d_g^v > 0$, otherwise the optimal solution (in all objective functions considered) would be the one without any delay, i.e., $y_{arr}{}_g^v = y_{dep}{}_g^v = 0$ for all $g \in F, v \in V^g$ and $\bar{z}_{ghv} = 0$ for all $(g,h,v) \in \mathcal{U}$.

The conditions (6.9) and (6.10) are rewritten to the y-variables in constraints (6.13) and (6.14), see again Figure 6.3. Finally, (6.15) enforces that also for all maintained connections the delay is carried over correctly, where it can be reduced by the slack time that can be saved if the passengers change as quickly as possible. The situation is depicted in part c) of Figure 6.3.

In the next section we present a more elegant way of stating the delay management problem. This approach is based on event-activity networks and handles the three types of activities (waiting, driving, changing) together with their slack times in a uniform way. To the best of our knowledge it has not been used for analyzing the delay management problem before, except in [Gin01, GS02].

6.4 Event-activity Networks in Delay Management

A more convenient way of describing the delay management problem is to represent the PTN as an activity-on-arc project network. In timetabling, such a concept has been used frequently, see, e.g., the PhD-theses [Wei81, Ste88, Kri96, Nac98, Pee02] and references therein. For our purpose of delay management, we propose the following representation of the PTN, which is based on [Nac98].

Notation 6.4. *Given a public transportation network* PTN $= (V, E)$, *a set of vehicles* F, *and a timetable specifying the departure and arrival times of each vehicle at each station, the corresponding* **event-activity network** $\mathcal{N} = (\mathcal{E}, \mathcal{A})$ *is constructed as follows. We define a set of nodes, representing* **arrival events** *and* **departure events**

$$\mathcal{E} = \mathcal{E}_{arr} \cup \mathcal{E}_{dep},$$

and a set of directed arcs, representing **driving activities**, **waiting activities**, *and* **changing activities**

$$\mathcal{A} = \mathcal{A}_{drive} \cup \mathcal{A}_{wait} \cup \mathcal{A}_{change}$$

as follows:

$$\begin{aligned}
\mathcal{E}_{arr} &= \{(g, v, \text{arr}) : \text{ vehicle } g \in F \text{ arrives at station } v \in V\}, \\
\mathcal{E}_{dep} &= \{(g, v, \text{dep}) : \text{ vehicle } g \in F \text{ departs from station } v \in V\}, \\
\mathcal{A}_{drive} &= \{((g, v, \text{dep}), (g, u, \text{arr})) \in \mathcal{E}_{dep} \times \mathcal{E}_{arr} : \text{ vehicle } g \text{ goes} \\
&\qquad \text{directly from station } v \text{ to } u\}, \\
\mathcal{A}_{wait} &= \{((g, v, \text{arr}), (g, v, \text{dep})) \in \mathcal{E}_{arr} \times \mathcal{E}_{dep}\}, \\
\mathcal{A}_{change} &= \{((g, v, \text{arr}), (h, v, \text{dep})) \in \mathcal{E}_{arr} \times \mathcal{E}_{dep} : \text{ a changing} \\
&\qquad \text{possibility from vehicle } g \text{ into } h \text{ at station } v \text{ should be provided}\}.
\end{aligned}$$

Furthermore, let $C = |\mathcal{A}_{change}|$.

The driving and waiting activities are performed by vehicles, while the changing activities are used by the customers. In short we can write

$$\mathcal{A}_{change} = \{((g, v, \text{arr}), (h, v, \text{dep})) : (g, h, v) \in \mathcal{U}\},$$

hence $\mathcal{U}$ can be identified with $\mathcal{A}_{change}$, especially $|\mathcal{U}| = C$.

An example of a PTN and its corresponding event-activity network is given in Figures 6.4 and 6.5.

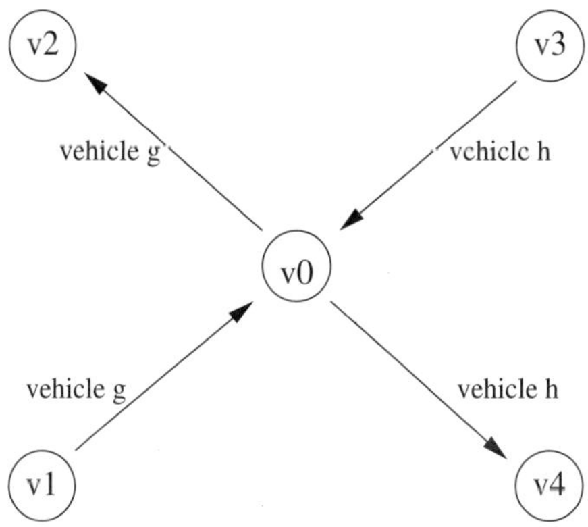

Fig. 6.4. A PTN with two vehicles g and h ...

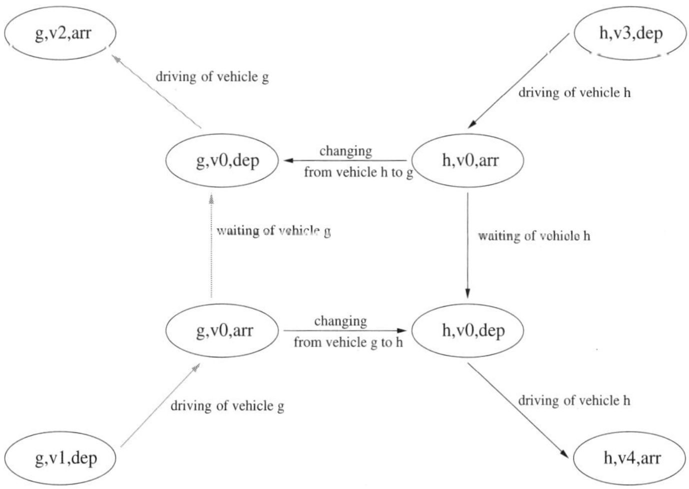

Fig. 6.5. ...and its corresponding event-activity network.

In the context of the delay management problem the event activity network according to Notation 6.4 is a special case of a time-expanded network, in which we can assume that a feasible timetable Π is known. Hence, $\mathcal{N}$ is acyclic, motivating the next definition.

Definition 6.5.

1. *Let $i, j \in \mathcal{E}$.*
   - $i \prec j$ *if there exists a path in $\mathcal{N}$ containing event i before j.*
   - $i \preceq j$ *if $i \prec j$ or $i = j$.*
2. *Let $a, a' \in \mathcal{A}$.*
   - $a \prec a'$ *if there exists a path in $\mathcal{N}$ containing activity a before a'.*
   - $a \preceq a'$ *if $a \prec a'$ or $a = a'$.*
3. *Let $I \subseteq \mathcal{E}$. Then i is a minimal element of I if there does not exist some $j \in I$ such that $j \prec i$.*
4. *Let $\mathcal{A}^0 \subseteq \mathcal{A}$. Then a is a minimal element of $\mathcal{A}^0$ if there does not exist some $a' \in \mathcal{A}^0$ such that $a \prec a'$.*

$\preceq$ defines a partial order (and $\prec$ a strict partial order) both on the set of events and on the set of activities. In particular we know that for all sets $I \subseteq \mathcal{E}$ (or $\mathcal{A}^0 \subseteq \mathcal{A}$) a minimal element exists, but it need not be unique. If and only if the set I (or $\mathcal{A}^0$) is contained in one path, $\prec$ leads to a unique order of the set.

Using the notation of event-activity networks, a *timetable* Π is given by assigning a time Π_i to each event $i \in \mathcal{E}$ (see [Nac98]). The planned duration of activity $a = (i, j)$ is hence given by $\Pi_j - \Pi_i$. Furthermore, let L_a be the minimal duration of activity a. Formally, the relation between L_a and the minimal durations L_g^v, L_g^{vu}, and L_{gh}^v introduced on page 100 is the following:

- If $a = (i, j)$ is a driving activity, where i represents the departure of some vehicle g at some station v and j represents the arrival of g at another station u, then $L_a = L_g^{vu}$.
- If $a = (i, j)$ is a waiting activity, where i represents the arrival of vehicle g at station v and j its departure at the same station, then $L_a = L_g^v$.
- Finally, if $a = (i, j)$ is a changing activity, say i represents the arrival of vehicle g at station v and j represents the departure of another vehicle h at the same station v, then $L_a = L_{gh}^v$.

In all three cases, the corresponding slack time s_a represents the time which can be saved while performing activity a as fast as possible, and it is given by

$$s_a = \Pi_i - \Pi_j - L_a$$

for *all* three types of activities $a = (i, j) \in \mathcal{A}$. This unifies the three different formulas (6.2), (6.4), and (6.6) introduced on page 100.

Definition 6.1 can hence be reformulated as follows.

A timetable $\Pi \in \mathbb{N}^{|\mathcal{E}|}$ is feasible if for all $a = (i,j) \in \mathcal{A}$,

$$\Pi_j - \Pi_i \geq L_a.$$

The next lemma relates the precedence relation with the timetable, and shows how a set $I \subseteq \mathcal{E}$ can be ordered according to $\prec$ if all minimal durations L_a are strictly greater than zero.

Lemma 6.6. *Let Π be a feasible timetable in $\mathcal{N}$.*

1. *If $i \prec j$ then $\Pi_i \leq \Pi_j$.*
2. *Let $I \subseteq \mathcal{E}$. If $\Pi_i = \min_{j \in I} \Pi_j$ and $L_a > 0$ for all $a = (j,i) \in \mathcal{A}$ with $j \in I$, then i is a minimal element of I with respect to $\prec$.*

Note that the reverse directions are in general wrong for the statements of Lemma 6.6. On the other hand, if $I = \{i_1, i_2, \ldots, i_{|I|}\} \subseteq \mathcal{E}$ is sorted according to the scheduled times, i.e.

$$\Pi_{i_1} \leq \Pi_{i_2} \cdots \leq \Pi_{i_{|I|}}$$

and $L_a > 0$ for all $a = (i,j) \in \mathcal{A}$ with $i, j \in I$ we obtain that $i_k \not\prec i_{k-1}$ for all $k = 2, \ldots, |I|$. A set of activities $\mathcal{A}^0 \subseteq \mathcal{A}$ can be ordered similarly.

As before, we assume that all *source delays* are known, i.e., we have a set of (arrival) events $\mathcal{E}_{del} \subseteq \mathcal{E}_{arr}$ such that $d_i > 0$ for all $i \in \mathcal{E}_{del}$. For non-delayed events we again set $d_i = 0$. This means that in case of a delay of (arrival) event $i \in \mathcal{E}_{arr}$, the actual arrival time of i does not coincide with the scheduled time Π_i. Increasing Π_i by d_i may lead to an infeasible timetable. As before the goal is to calculate a perturbed feasible timetable x_i for all events $i \in \mathcal{E}$. Due to its importance for the remainder of this part, we repeat the definition of a *feasible perturbed timetable* on page 102 in our new notation.

Definition 6.7. *A* **perturbed timetable** *x_i for all $i \in \mathcal{E}$ is* **feasible**, *if the following conditions hold.*

$$x_i \geq \Pi_i + d_i \text{ for all } i \in \mathcal{E} \text{ and} \tag{6.16}$$
$$x_j - x_i \geq L_a \text{ for all } a = (i,j) \in \mathcal{A}_{wait} \cup \mathcal{A}_{drive}. \tag{6.17}$$

Constraint (6.16) requires that no event must be scheduled earlier than in the original timetable, and furthermore ensures that for all $i \in \mathcal{E}_{del}$ the source delays are taken into account. Due to constraint (6.17) the delay is carried over correctly from one event to the next one along waiting and driving activities. We remark that this definition of feasibility is equivalent to Definition 6.2.

We also rewrite the integer programming formulation of feasible solutions. We now use variables with the following meaning:

$$y_i = \text{delay of event } i$$
$$z_a = \begin{cases} 0 & \text{if changing activity } a \in \mathcal{A}_{change} \text{ is maintained} \\ 1 & \text{if changing activity } a \in \mathcal{A}_{change} \text{ is missed.} \end{cases}$$

Then $(y, \bar{z}) \in \mathbb{N}^{|\mathcal{E}|} \times \{0,1\}^C$ is a feasible solution of the delay management problem, if

$$y_i \geq d_i \quad \text{for all } i \in \mathcal{E}_{del} \tag{6.18}$$
$$y_i - y_j \leq s_a \quad \text{for all } a = (i,j) \in \mathcal{A}_{wait} \cup \mathcal{A}_{drive} \tag{6.19}$$
$$-M\bar{z}_a + y_i - y_j \leq s_a \quad \text{for all } a = (i,j) \in \mathcal{A}_{change}. \tag{6.20}$$

We denote

$$\text{Feas}_{DM} = \{(y, \bar{z}) \in \mathbb{N}^{|\mathcal{E}|} \times \{0,1\}^C : (y, \bar{z}) \text{ satisfies (6.18), (6.19), and (6.20)}\}$$

as the set of all feasible solutions of the delay management problem.

The necessary size of M in (6.20) depends on the objective function used. If we know that in all solutions which we are interested in $y_i \leq H$ for any fixed value H, then $M \geq H$ is large enough, since it allows that connections are **not** maintained.

We remark that the second notation is more general than the "intuitive" notation of the previous section since it can be applied to any directed acyclic network $\bar{\mathcal{N}}$, while the event-activity network $\mathcal{N}$ corresponding to a timetable is a very special case of such a graph. (E.g., for each node in $\mathcal{N}$ either the in-degree or the out-degree has to be 1.) Since many models and procedures which will be presented in the following do not rely on the special structure of the event-activity network $\mathcal{N}$ they can be applied to more general problems in any acyclic directed network. The question is if our problems become harder by forgetting about the special structure of $\mathcal{N}$. This is (in terms of complexity) not the case, since it can be shown that *each* acyclic directed graph $\bar{\mathcal{N}}$ can be transformed to the event-activity network of some appropriate PTN (according to Notation 6.4) with the same precedence structure in polynomial time.

7

Delay Management With Fixed Connections

In this chapter we focus on how to calculate a perturbed feasible timetable which is as punctual as possible, if a set of connections to be maintained has already been fixed. The goal of this chapter is to find "tight" timetables, i.e. timetables which minimize the (weighted) sum of differences to the given timetable and maintain the fixed connections. We will show that both,

- for minimizing the total delay over all customers (TDM, Chapter 8), and
- for finding efficient solutions in the bicriteria problem (BDM, Chapter 9),

we can restrict our search to such "tight" timetables. A proper description of a "tight" timetable will be given by the concept of a *time-minimal* solution in Notation 7.7. The problem of this chapter is the following.

(TT($\mathcal{U}^{fix}$))

Given PTN, $F, \mathcal{U}$, *minimal necessary times for driving, waiting, and changing, a feasible timetable* Π_{arr}, Π_{dep}, *a set of delayed events* $\mathcal{E}_{del}$, *a set of weights* ${w^{fix}}^v_g$ *for all vehicles at all stations, and a set of connections* $\mathcal{U}^{fix} \subseteq \mathcal{U}$, *find a perturbed feasible timetable* x_{arr}, x_{dep}, *maintaining all connections in* $\mathcal{U}^{fix}$ *such that*

$$f_{TT} = \sum_{g \in F, v \in V^g} {w^{fix}}^v_g ({x_{arr}}^v_g - {\Pi_{arr}}^v_g)$$

is minimized.

Chapter 7 is structured as follows: (TT) can be interpreted as finding a (non-periodic) timetable. In this chapter we present three solution approaches to solve (TT).

- Our first approach is based on an integer programming formulation for (TT). We show that its coefficient matrix is totally unimodular.

- The second approach uses the *critical path method* (CPM) of project planning, and
- the third one clarifies the relation to the feasible differential problem.

Investigating the structure of the optimal solution we prove some properties that will be needed frequently in the following chapters.

7.1 Linear Programming Approach

In the notation of event-activity networks, the set of connections to be maintained is denoted by $\mathcal{A}^{fix} \subseteq \mathcal{A}_{change}$. To derive an integer programming formulation we can use the formulation of Feas_{DM} (see page 108), but fix $\bar{z}_a = 0$ for all $a \in \mathcal{A}^{fix}$. The remaining variables $\bar{z}_a$, $a \notin \mathcal{A}^{fix}$ are not known beforehand. The condition $y_i \in \mathbb{N}$ in the definition of Feas_{DM} is not needed. This will become clear in Theorem 7.1 below. Consequently, we obtain the following linear program.

(TT($\mathcal{A}^{fix}$))

$$\min f_{\text{TT}} = \sum_{i \in \mathcal{E}} w_i^{fix} y_i$$

such that

$$y_i \geq d_i \quad \forall i \in \mathcal{E} \tag{7.1}$$

$$y_i - y_j \leq s_a \quad \forall a = (i,j) \in \mathcal{A}_{wait} \cup \mathcal{A}_{drive} \cup \mathcal{A}^{fix}. \tag{7.2}$$

We now justify that the constraints $y_i \in \mathbb{N}$ are in fact not necessary for $\text{TT}(\mathcal{A}^{fix})$.

Theorem 7.1. *All extreme points of the feasible set of TT($\mathcal{A}^{fix}$) are integer.*

Proof. The coefficient matrix of $\text{TT}(\mathcal{A}^{fix})$ is given by

$$\Phi = \begin{pmatrix} -I_{|\mathcal{E}|} \\ \Theta^T \end{pmatrix},$$

where $I_{|\mathcal{E}|}$ denotes the $|\mathcal{E}| \times |\mathcal{E}|$ unit matrix, and Θ is the node-arc incidence matrix of $\mathcal{N}$, i.e.,

$$\Theta_{i,a} = \begin{cases} -1 & \text{if } a = (i,j) \text{ for some } j \in \mathcal{E} \\ 1 & \text{if } a = (j,i) \text{ for some } j \in \mathcal{E} \\ 0 & \text{otherwise.} \end{cases}$$

Since Θ is a totally unimodular matrix, Θ^T and hence Φ is. Together with $s_a \in \mathbb{N}$ this yields that all basic solutions of the linear programming relaxation are integer, see Appendix A. □

This means that for given connections $\bar{z}$ to be maintained the delay y can be assumed to be in minutes as required, and $(\mathrm{TT}(\mathcal{A}^{fix}))$ can be solved by linear programming. Other solution approaches that supply more structural insight are described next. Since these methods are more intuitive if we deal with timetables, we rewrite $(\mathrm{TT}(\mathcal{A}^{fix}))$ using the variables $x_i, i \in \mathcal{E}$ describing the perturbed timetable instead of the delay. This means we replace y_i by $x_i - \Pi_i$. We equivalently obtain

$(\mathbf{TT}^x(\mathcal{A}^{fix}))$

$$\min f_{\mathrm{TT}} = \sum_{i \in \mathcal{E}} w_i^{fix} x_i$$

such that

$$x_i \geq \Pi_i + d_i \quad \forall i \in \mathcal{E} \tag{7.3}$$

$$x_j - x_i \geq L_a \quad \forall a = (i,j) \in \mathcal{A}_{wait} \cup \mathcal{A}_{drive} \cup \mathcal{A}^{fix}. \tag{7.4}$$

Finally, let us simplify the notation as follows.

Notation 7.2. *Given a set $\mathcal{A}^{fix} \subseteq \mathcal{A}_{change}$ define*

$$\mathcal{A}(\mathcal{A}^{fix}) = \mathcal{A}_{wait} \cup \mathcal{A}_{drive} \cup \mathcal{A}^{fix}.$$

7.2 Relation to the Critical Path Method

Definition 7.3 (e.g.,[Elm77]). *A* **project network** *$\bar{\mathcal{N}} = (\bar{\mathcal{E}}, \bar{\mathcal{A}})$ is an acyclic digraph with exactly one source s and one sink t, such that for each event $i \in \bar{\mathcal{E}}$ there exists an s-t-path in $\bar{\mathcal{N}}$ containing i. Furthermore, each activity $a \in \bar{\mathcal{A}}$ has been assigned a minimal duration $\bar{L}_a \geq 0$.*

Since project networks are acyclic, the precedence relation $\prec$ defines a partial order of the nodes $\bar{\mathcal{E}}$, given by $a' \prec a$ for $a, a' \in \bar{\mathcal{A}}$, if a' occurs before a on a path from s to t (see also Definition 6.5). Activity a can only be performed, if all activities a' with $a' \prec a$ have already been completed. The goal in *project planning* is to find the minimal necessary completion time of the whole project, i.e., the earliest possible time point when arriving at the sink t. The starting time for performing the first activity within the project network is known, and assumed to be 0. The minimal necessary completion time can be determined, e.g., by the critical path method (CPM). This well-known procedure (see, e.g., [Elm77, MM98]) consists of two phases, a forward and a backward phase, where the forward phase works as follows:

(CPM-F) Forward phase of CPM: For each event i the earliest possible starting time x_i is determined iteratively by

- $x_s = 0$
- $x_j = \max_{a=(i,j)\in\bar{\mathcal{A}}}\{x_i + \bar{L}_a\}$.

Note that x_t then gives the smallest possible completion time of the project.

To apply the forward phase of CPM for solving $\mathrm{TT}(\mathcal{A}^{fix})$ we construct the following network $\mathcal{N}^s(\mathcal{A}^{fix})$ from the given event-activity network $\mathcal{N}$. Note that a sink is not necessary if only the forward phase of CPM is applied.

Notation 7.4. *Let $\mathcal{N} = (\mathcal{E}, \mathcal{A})$ and let $\mathcal{A}^{fix} \subseteq \mathcal{A}_{change}$. The corresponding* **CPM-network** $\mathcal{N}^s(\mathcal{A}^{fix}) = (\mathcal{E}^s, \mathcal{A}^s(\mathcal{A}^{fix}))$ *is given by*

$$\mathcal{E}^s = \mathcal{E} \cup \{s\}$$
$$\mathcal{A}^s(\mathcal{A}^{fix}) = \mathcal{A}_{wait} \cup \mathcal{A}_{drive} \cup \mathcal{A}^{fix} \cup \{(s,i) : i \in \mathcal{E}\}$$
$$L_a^x = \begin{cases} L_a & \text{if } a \in \mathcal{A}_{wait} \cup \mathcal{A}_{drive} \cup \mathcal{A}^{fix} \\ \Pi_i + d_i & \text{if } a = (s,i), i \in \mathcal{E}. \end{cases}$$

The additional arcs $a = (s,i) \in \mathcal{A}^s \setminus \mathcal{A}$ are called **timetable arcs.**

The timetable arcs represent the given timetable, and the known source delays. They make sure that no vehicle departs earlier than originally planned, and that the source delays are taken into account. Applying the forward phase of CPM to the CPM-network $\mathcal{N}^s$ yields a timetable x_i for all $i \in \mathcal{E}$. We have the following result.

Theorem 7.5. *Let $\bar{x}$ be a timetable calculated by* (CPM-F). *Then $\bar{x}_i, i \in \mathcal{E}$ is an optimal solution of ($TT(\mathcal{A}^{fix})$).*

Proof. Since $\bar{x}$ is the timetable calculated by (CPM-F) we have

$$x_s = 0 \text{ and}$$
$$\bar{x}_j \geq \bar{x}_i + L_a^x \text{ for all } a = (i,j) \in \mathcal{A}^s(\mathcal{A}^{fix}).$$

This means,

- $\bar{x}_j \geq \bar{x}_i + L_a$ for all $a = (i,j) \in \mathcal{A}$, and
- $\bar{x}_j \geq \Pi_j + d_j$ for all $a = (s,j)$,

hence $\bar{x}$ is feasible for $(\mathrm{TT}^x(\mathcal{A}^{fix}))$. Now let x be another feasible timetable such that

$$f_{\mathrm{TT}}(x) = \sum_{i\in\mathcal{E}} w_i^{fix} x_i < \sum_{i\in\mathcal{E}} w_i^{fix} \bar{x}_i = f_{\mathrm{TT}}(\bar{x}).$$

Let j^* be a minimal event, where the timetable calculated by (CPM-F) is later than x, i.e., take j^* minimal w.r.t. $\prec$ satisfying $x_{j^*} < \bar{x}_{j^*}$. Then

$$\begin{aligned} x_{j^*} < \bar{x}_{j^*} &= \max_{a=(i,j^*)\in\mathcal{A}^s} \bar{x}_i + L^x_{(i,j^*)} \\ &\leq \max_{a=(i,j^*)\in\mathcal{A}^s} x_i + L^x_{(i,j^*)} \\ &= x_{i^*} + L^x_{(i^*j^*)} \text{ for some } i^* \in \mathcal{E}. \end{aligned}$$

Case 1: $i^* \in \mathcal{E}$. Then $x_{j^*} - x_{i^*} < L^x_{(i^*j^*)}$ contradicts the feasibility of x, see constraint (7.4).

Case 2: $i^* = s$. Then $x_{i^*} = \bar{x}_{i^*} = 0$ and $L^x_{i^*j^*} = \Pi_{j^*} + d_{j^*}$, meaning that $x_{j^*} < \Pi_{j^*} + d_{j^*}$ contradicting constraint (7.3) and hence again the feasibility of x. □

Corollary 7.6. *Let $\mathcal{A}^{fix} \subseteq \mathcal{A}_{change}$. Then there exists an optimal solution of $(TT(\mathcal{A}^{fix}))$ satisfying*

$$y_j = \max\{d_j, \max_{a=(i,j)\in\mathcal{A}(\mathcal{A}^{fix})} y_i - s_a\}$$

for all $j \in \mathcal{E}$.

Proof. Let x be an optimal solution for TT($\mathcal{A}^{fix}$) constructed by (CPM-F) in $\mathcal{N}^s(\mathcal{A}^{fix})$. Let $j \in \mathcal{E}$. Then according to (CPM-F),

$$\begin{aligned} x_s &= 0 \\ x_j &= \max_{a=(i,j)\in\mathcal{A}^s} x_i + L^x_a \\ &= \max\{\Pi_j + d_j, \max_{a=(i,j)\in\mathcal{A}(\mathcal{A}^{fix})} x_i + L_a\}, \end{aligned}$$

where for the second equation we used the definition of $\mathcal{A}^s(\mathcal{A}^{fix})$ and of L^x_a according to Notation 7.4. Replacing x_j by $y_j + \Pi_j$ and using that $\Pi_j - \Pi_i - L_a = s_a$ finally yields

$$\begin{aligned} y_j &= x_j - \Pi_j \\ &= \max\{\Pi_j + d_j, \max_{a=(i,j)\in\mathcal{A}(\mathcal{A}^{fix})} y_i + \Pi_i + L_a\} - \Pi_j \\ &= \max\{d_j, \max_{a=(i,j)\in\mathcal{A}(\mathcal{A}^{fix})} y_i - s_a\}. \end{aligned}$$

□

We remark that y can also directly be found by applying (CPM-F) in $\mathcal{N}^s(\mathcal{A}^{fix}) = (\mathcal{E}^s, \mathcal{A}^s(\mathcal{A}^{fix}))$ with

$$L^y_a = \begin{cases} -s_a & \text{if } a \in \mathcal{A}_{wait} \cup \mathcal{A}_{drive} \cup \mathcal{A}^{fix} \\ d_i & \text{if } a = (s, i), i \in \mathcal{E}. \end{cases} \tag{7.5}$$

Notation 7.7. *Let $\mathcal{A}^{fix} \subseteq \mathcal{A}_{change}$. An optimal solution for $(TT(\mathcal{A}^{fix}))$ satisfying the properties of Corollary 7.6 is called a* **time-minimal** *solution, and is denoted by $y(\mathcal{A}^{fix})$.*

Due to Corollary 7.6 a time-minimal solution can be calculated by the following algorithm.

Algorithm 12: Calculating a time-minimal solution for TT($\mathcal{A}^{fix}$)

```
Input: 𝒩, d_i, s_a, A^fix.
Output: Optimal (time-minimal) solution of TT(A^fix).
Step 1. Sort ℰ = {i_1,...,i_|ℰ|} according to ≺.
Step 2. For k = 1,...,|ℰ|: y_{i_k} = max{d_{i_k}, max_{a=(i,i_k)∈A(A^fix)} y_i − s_a}
Step 3. Output: y_i, i ∈ ℰ
```

We conclude this section by presenting some properties which we will use in the subsequent chapters about delay management.

Lemma 7.8. *Let $\mathcal{A}^1, \mathcal{A}^2 \subseteq \mathcal{A}_{change}$ be two sets of connections which have to be maintained. Then the following hold.*

1. *$y(\mathcal{A}^1) \leq y$ for all feasible solutions y of $(TT(\mathcal{A}^1))$.*
2. *If $\mathcal{A}^1 \subseteq \mathcal{A}^2$ then $y(\mathcal{A}^1) \leq y(\mathcal{A}^2)$.*

Proof.

1. The first part is shown by induction.
Start: Let i be a minimal element of $\mathcal{E}$. Then $y_i^* = d_i \leq y_i$.
Conclusion: Let the induction hypothesis be true for all $j \prec i$. Then

$$\begin{aligned} y_i^* &= \max\{d_i, \max_{a=(j,i)\in\mathcal{A}(\mathcal{A}^1)} y_j^* - s_a\} \\ &\leq \max\{d_i, \max_{a=(j,i)\in\mathcal{A}(\mathcal{A}^1)} y_j - s_a\} \leq y_i, \end{aligned}$$

due to (7.1) and (7.2) on page 110.
2. Note that (TT($\mathcal{A}^1$)) is a relaxation of (TT($\mathcal{A}^2$)), since the only difference between both problems is that in the latter we have possibly more constraints of type (7.2). Consequently, $y(\mathcal{A}^2)$ is feasible for TT($\mathcal{A}^1$), and the result follows according to part 1. □

The last property is an immediate consequence of Lemma 7.8 and justifies the small size of M in the integer programming formulation of Feas$_{DM}$.

Corollary 7.9. *Let $\mathcal{A}^{fix} \subseteq \mathcal{A}_{change}$, let y^* be the time-minimal solution of $(TT(\mathcal{A}^{fix}))$ and let $D = \max_{i\in\mathcal{E}_{del}} d_i$. Then $y_i^* \leq D$ for all $i \in \mathcal{E}$.*

Proof. It can be easily verified that $y_i = D$ is a feasible solution of (TT($\mathcal{A}^{fix}$)). Hence the result follows from Lemma 7.8, part 1. □

7.3 Relation to the Feasible Differential Problem

For an instance of the feasible differential problem (see, e.g., [Roc84]) we need

- a directed, connected network, which will in the following be denoted by $\bar{\mathcal{N}} = (\bar{\mathcal{E}}, \bar{\mathcal{A}})$ and
- upper and lower bounds $\bar{L}_a$ and $\bar{U}_a$ on each arc $a \in \bar{\mathcal{A}}$.

The *feasible differential problem* consists of finding a node potential $\bar{x}_i$ for each node $i \in \bar{\mathcal{E}}$ such that for all arcs $a = (i, j) \in \bar{\mathcal{A}}$:

$$\bar{L}_a \leq \bar{x}_i - \bar{x}_j \leq \bar{U}_a.$$

Notation 7.10.

- *For a node $i \in \mathcal{E}$, $\bar{x}_i$ is called the* **potential** *of i.*
- *For an arc $a = (i, j) \in \bar{\mathcal{A}}$, $\bar{t}_a = \bar{x}_j - \bar{x}_i$ is called the* **tension** *of a with respect to $\bar{x}$.*
- *A vector $\bar{t} \in \mathbb{R}^{|\bar{\mathcal{A}}|}$ is called* **tension** *of $\bar{\mathcal{N}}$ if there exists a potential $\bar{x}$ such that $\bar{t}$ is the tension with respect to $\bar{x}$.*

Let the $|\bar{\mathcal{E}}| \times |\bar{\mathcal{A}}|$-matrix $\bar{\Theta}$ denote the node-arc-incidence matrix of $\bar{\mathcal{N}}$. This means,

$$\bar{\Theta}_{i,a} = \begin{cases} -1 & \text{if } a = (i, j) \text{ for some } j \in \bar{\mathcal{E}} \\ 1 & \text{if } a = (j, i) \text{ for some } j \in \bar{\mathcal{E}} \\ 0 & \text{otherwise.} \end{cases}$$

Then

$$\bar{t} = \bar{\Theta}^T \bar{x}.$$

The following observation is obvious but will be helpful later.

Lemma 7.11. *Let p be any path from i to j in $\bar{\mathcal{N}}$. Then*

$$\sum_{a \in p} \bar{t}_a = \bar{x}_j - \bar{x}_i.$$

Proof. Let p be represented by its events $(i = i_0, i_1, i_2, \ldots, i_P = j)$. Then

$$\sum_{a \in p} \bar{t}_a = \sum_{k=1}^{P} \bar{x}_{i_k} - \bar{x}_{i_{k-1}} = \bar{x}_{i_P} - \bar{x}_{i_0} = \bar{x}_j - \bar{x}_i.$$

□

Furthermore, let the $\left(|\bar{\mathcal{A}}| - |\bar{\mathcal{E}}| + 1\right) \times |\bar{\mathcal{A}}|$-matrix $\bar{\Gamma}$ be a network matrix of $\bar{\mathcal{N}}$. Such a network matrix can be constructed as follows:
Let $\bar{\mathcal{T}}$ be a spanning tree in $\bar{\mathcal{N}}$. Then, adding some $\tilde{a} \in \bar{\mathcal{A}} \setminus \bar{\mathcal{T}}$ gives a circuit $C_{\tilde{a}}$ in $\bar{\mathcal{N}}$. The entries of $\bar{\Gamma}$ are then given by:

$$\bar{\Gamma}_{\tilde{a},a} = \begin{cases} 1 & \text{if } C_{\tilde{a}} \text{ contains } a \text{ and } \tilde{a} \text{ in the same direction} \\ -1 & \text{if } C_{\tilde{a}} \text{ contains } a \text{ and } \tilde{a} \text{ in opposite directions} \\ 0 & \text{if } C_{\tilde{a}} \text{ does not contain } a. \end{cases}$$

An interesting question is to determine, if a given vector $\bar{t}$ is a tension in $\bar{\mathcal{N}}$, i.e., if there exists some potential $\bar{x}$ such that $\bar{t}$ is the tension with respect to $\bar{x}$. From Lemma 7.11 we directly get a necessary condition: If $\bar{t}$ is a tension, then it has to satisfy

$$\sum_{a \in C} \bar{t}_a = 0$$

for any circuit C in $\bar{\mathcal{N}}$. The next theorem shows that this condition is

a) sufficient, and
b) equivalent to the requirement that

$$\sum_{a \in C_{\tilde{a}}} \bar{t}_a = 0$$

for all circuits $C_{\tilde{a}}$ appearing in the rows of (any) network matrix.

Theorem 7.12 (e.g., [Elm77]). *Let $\bar{\Gamma}$ be a network matrix in $\bar{\mathcal{N}}$. Then $\bar{t}$ is a tension in $\bar{\mathcal{N}}$ if and only if*

$$\bar{\Gamma}\bar{t} = 0,$$

where 0 is the zero vector with $|\bar{\mathcal{A}}| - |\bar{\mathcal{E}}| + 1$ components.

Now we are going to apply these results for (TT($\mathcal{A}^{fix}$)). To this end, let Θ be the node-arc-incidence matrix of the event-activity network $\mathcal{N}$. As potential $\bar{x}$ we can use either the perturbed timetable x or the delay y. Then t^x denotes the tension in $\mathcal{N}$ with respect to x if $t^x = \Theta^T x$, and t^y is the tension with respect to the delay if $t^y = \Theta^T y$. To model (TT($\mathcal{A}^{fix}$)) we again use the CPM-network $\mathcal{N}^s(\mathcal{A}^{fix})$, constructed in Notation 7.4 (page 112). Neglecting any upper bound, we hence get a feasible differential problem in an acyclic graph, which is equivalent to (TT^x($\mathcal{A}^{fix}$)). Due to Theorem 7.12 its formulation for t^x is given by

(TT^x($\mathcal{A}^{fix}$)-tension)

$$\min \sum_{a=(s,i):i \in \mathcal{E}} t_a^x w_a^{fix}$$

such that

$$t_a^x \geq L_a^x$$
$$\Gamma^s t^x = 0,$$

where Γ^s is a network matrix of $\mathcal{N}^s$, and L_a^x is given as in Notation 7.4. From $x = \Pi + y$ we conclude that $t^x = \Theta^T \Pi + t^y$. Equivalently, using y as potential we obtain

(TT^y($\mathcal{A}^{fix}$)-tension)

$$\min \sum_{a=(s,i):i\in\mathcal{E}} t_a^y w_a^{fix}$$

such that

$$\begin{aligned} t_a^y &\geq L_a^y \\ \Gamma^s t^y &= 0, \end{aligned}$$

where L_a^y is given as in (7.5).

Some remarks are added.

- Given a tension t^x (t^y), a corresponding node potential x (y) can be constructed by setting $x_s = 0$ ($y_s = 0$) and calculating x_i (y_i) as the length of (any) longest path from s to i with respect to L_a^x (L_a^y). Note that this length is the same for all paths from s to i due to Theorem 7.12.
- If the network $\mathcal{N}^s$ does not contain any circuit, then the constraint $\Gamma^s t^x = 0$ ($\Gamma^s t^y = 0$) can be omitted, and the feasible differential problem is trivially solvable by setting $t_a^x = \bar{L}_a^x$ ($t_a^y = \bar{L}_a^y$) for all $a \in \bar{\mathcal{A}}$.
- To construct a network matrix Γ^s we can proceed as follows. As spanning tree $\mathcal{T} \subseteq \mathcal{A}^o$ we take
$$\mathcal{T} = \{(s,i) : i \in \mathcal{E}\}.$$
Note that $\mathcal{A}^s \setminus \mathcal{T} = \mathcal{A}$, hence the corresponding network matrix Γ^s has $|\mathcal{A}|$ rows and $|\mathcal{A}| + |\mathcal{E}|$ columns and is given by
$$\Gamma^s = \left(I_{|\mathcal{A}|} \; , \; -\Theta^T \right).$$

In the following we show that the longest path technique for solving the feasible differential problem can be applied to the delay management problem with fixed connections. First, we focus on finding a perturbed timetable x.

Lemma 7.13. *Let $x_i^* = LoP(s,i)$ be the length of a longest path from s to i in $\mathcal{N}^s(\mathcal{A}^{fix})$ with weights L^x. Then x^* is an optimal solution of (TT^x($\mathcal{A}^{fix}$)).*

Proof. First, x^* is feasible, since

- $x_i^* = LoP(s,i) \geq L_{(s,i)}^x = \Pi_i + d_i$, and
- $LoP(s,j) \geq LoP(s,i) + L_{(i,j)}^x$ for any $a = (i,j) \in \mathcal{A}$, hence
$$x_j^* - x_i^* \geq L_a.$$

Now let x be any other feasible solution of $(\mathrm{TT}^x(\mathcal{A}^{fix}))$, and assume that $x_i < LoP(s,i)$ for some $i \in \mathcal{E}$. Choose i minimal with respect to $\prec$ with this property. Let p be a longest path from s to i, say with last edge $a = (\bar{\imath}, i) \in \mathcal{A}^s$. Consequently,

$$x_i < LoP(s,i) = LoP(s,\bar{\imath}) + L_a^x \leq x_{\bar{\imath}} + L_a^x,$$

which is a contradiction to $x_i \geq \Pi_i + d_i$ if $\bar{\imath} = s$, and to $x_i \geq x_{\bar{\imath}} + L_a$ if $\bar{\imath} \in \mathcal{E}$. □

We remark that y_i can also be calculated directly as longest path in $\mathcal{N}^s(\mathcal{A}^{fix})$ with weights L^y, and that the obtained solution equals $y(\mathcal{A}^{fix})$ of (CPM-F).

Corollary 7.14. *Let $y_i = LoP(s,i)$ be the length of a longest path from s to i in $\mathcal{N}^s(\mathcal{A}^{fix})$ with weights L^y. Then y is an optimal solution of $(TT(\mathcal{A}^{fix}))$.*

8

Minimizing the Sum of All Delays

In this chapter we will deal with the **total delay management problem (TDM)**, where we try to minimize the *total delay* defined as sum of all delays over all customers traveling through PTN. To deal with (TDM) we introduce two more assumptions.

- We assume that T is the (common) time period for all vehicles, and that
- in the next time period all vehicles are on time.

The first assumption is often made, especially in literature about timetabling. If T is taken as the largest time period over all lines, it overestimates the inconvenience for the customers. We mention that this assumption can be relaxed easily in the formulation (TDM-C) which will be presented in Section 8.2, but is needed for the linear model (TDM-B) in the next section.

The second assumption is no restriction in on-line decisions, since in the on-line case the planning horizon usually is not larger than the common time period T. If we want to plan for a complete day, this assumption only is reasonable, if there is enough slack time for the vehicles at the last stations of their respective lines.

We now need to specify the data about the customers. To this end, let $\mathcal{P}$ be a set of paths through the public transportation network PTN, which customers would like to use during their trips. For each path we need to specify not only the edges $e \in E$ of the path, but also a vehicle $g \in F$ for each edge. A path is hence described by a sequence

$$p = (v_1, g_1, v_2, g_2, \dots, g_{l_p}, v_{l_p+1}),$$

where $g_j \in F$ and $v_j \in V$ satisfy that $(v_j, v_{j+1}) \in E^{g_j}$ for all $j = 1, \dots, l_p$. Moreover, we assume that the customers using path p arrive at their first station v_1 on time by some means of transport not relevant for our system. For each path we introduce some weight $w_p > 0$ which gives the number of customers who wish to use path p. The following notation will be used when dealing with the paths. (For an illustration, see Figure 8.1.)

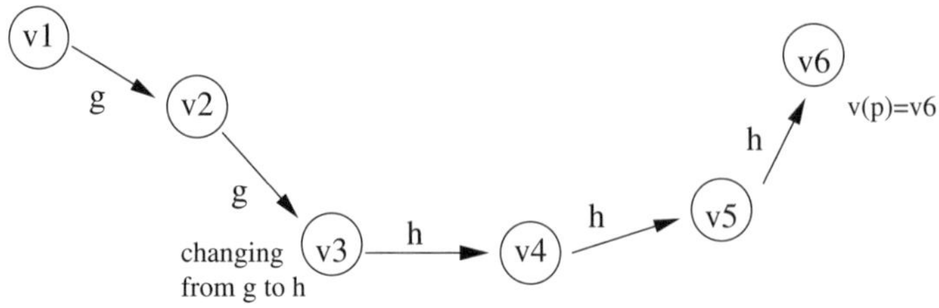

Fig. 8.1. A path p containing the connection (g, h, v_3).

Notation 8.1.

- *For the sake of simplicity we write* $(g, h, v) \in p$ *if we want to say that a connection* $(g, h, v) \in \mathcal{U}$ *is used in a path* $p \in \mathcal{P}$. *(Formally, if the sequence* (g, v, h) *is contained in* p.*)*
- *A path is called* **maintained** *if all connections* $(g, h, v) \in p$ *are maintained, otherwise the path is called* **missed**.
- *For a path* $p \in \mathcal{P}$ *let* $g(p)$ *denote the last vehicle used on path* p *and* $v(p)$ *be the destination station of path* p.

Without loss of generality let us assume that each customer's path $p \in \mathcal{P}$ contains a station at most once. Now consider passengers traveling along a path $p \in \mathcal{P}$ through the public transportation network.

Case 1: If all connections on path p are maintained, the delay of these passengers equals the arrival delay of their last vehicle $g(p)$ at their destination station $v(p)$, and is hence given by $(x_{arr}{}_{g(p)}^{v(p)} - \Pi_{arr}{}_{g(p)}^{v(p)})$.

Case 2: If at least one connection on path p is missed, we assume that the passengers using path p have to wait the whole time period T for the next vehicle going towards their destination, and their delay hence is T.

We now can state the model for minimizing the total delay.

(TDM)

Given PTN, $F, \mathcal{U}$, *minimal necessary times for driving, waiting, and changing, a feasible timetable* Π_{arr}, Π_{dep}, *a set of paths* $\mathcal{P}$ *through* PTN, *with weights* w_p *for all* $p \in \mathcal{P}$ *and a set of delayed events* $\mathcal{E}_{del}$, *find a perturbed feasible timetable* x_{arr}, x_{dep}, *such that*

$$f_{\text{TDM}} = \sum_{p \in \mathcal{P}: p \text{ is maintained}} w_p \left(x_{arr}{}_{g(p)}^{v(p)} - \Pi_{arr}{}_{g(p)}^{v(p)} \right) + \sum_{p \in \mathcal{P}: p \text{ is missed}} T w_p$$

is minimized.

As before, we may assume that source delays only occur at arrival events. Note that the NP-completeness of (TDM) has recently been shown in [GJPS05].

Chapter 8 is structured as follows: We first present the formulation (TDM-A) which can be linearized to (TDM-B). It is based on looking at each customer's path as a whole. Some structural results about both formulations are mentioned, e.g., that we can assume that the optimal solution is time-minimal according to Notation 7.7. Another approach to tackle (TDM) is presented in Section 8.2. Here we add up all delays over all activities to calculate the total delay, leading to an alternative integer programming formulation (TDM-C). We show that this model is equivalent to (TDM-A) and (TDM-B). Since (TDM-C) is rather complicated we simplify the model by fixing some parameters. Fortunately, this new problem has some nice properties:

- it is correct if the so-called *never-meet-property* holds, and
- in this case it can be solved in linear time.

Moreover, we show that it is easy to check whether the never-meet-property holds and that it is often almost satisfied in practice. For the general case we derive heuristics, lower bounds, and finally a branch and bound approach.

8.1 A Linear Model

Now let us use the more convenient notation introduced in Section 6.4. We hence represent the customer's paths by events, i.e.,

$$p = (i_1, i_2, \dots, i_{p_L})$$

where $i_k \in \mathcal{E}$ are events, and $(i_k, i_{k+1}) \in \mathcal{A}$ are activities. In each path we have that i_1 is a departure event, i_2 an arrival event, $i_3 \in \mathcal{E}_{dep}$ and so on until we reach the last event i_{p_L} which is again an arrival event. The following notation is the same as before, namely:

- $i(p)$ denotes the last event on path p, i.e., the arrival of the last vehicle used at the final station of path p, and
- $a \in p$ for some activity $a \in \mathcal{A}$ and some path $p \in \mathcal{P}$, if activity $a = (i, j)$ is performed on path p, i.e., there exists k such that $i = i_k, j = i_{k+1}$ are the kth and (k+1)th element of path p.

Again, consider passengers traveling along a path $p \in \mathcal{P}$ through the public transportation network, represented as event-activity network $\mathcal{N}$.

Case 1: If all connections on path p are maintained, the delay of these passengers equals the arrival delay of their last event $i(p)$, i.e., the arrival of their last vehicle at their destination station, and is hence given by $y_{i(p)}$.

Case 2: If at least one connection on path p is missed, the delay of the customers on path p is given by T, as before.

Suppose that some source delays d_i of some of the arrival events are given as an amount of time (in minutes). Let $D = \max\{d_i : i \in \mathcal{E}\}$ denote the largest of these source delays. We assume that $0 < D < T$.

To describe the feasible set of (TDM) we again use Feas_{DM} (see page 108). However, to keep track of the paths which are needed in the objective function f_{TDM} we replace the variables $\bar{z}_a$ by variables z_p for all $p \in \mathcal{P}$ with the following meaning.

$$z_p = \begin{cases} 0 & \text{if all connections on path } p \text{ are maintained} \\ 1 & \text{otherwise.} \end{cases}$$

Our first integer programming formulation of (TDM) can now be presented.

(TDM-A)

$$\min f_{\text{TDM-A}} = \sum_{p \in \mathcal{P}} w_p (y_{i(p)}(1 - z_p) + T z_p)$$

such that

$$\begin{aligned} y_i &\geq d_i \quad \text{for all } i \in \mathcal{E}_{del} && (8.1) \\ y_i - y_j &\leq s_a \quad \text{for all } a = (i,j) \in \mathcal{A}_{wait} \cup \mathcal{A}_{drive} && (8.2) \\ -M z_p + y_i - y_j &\leq s_a \quad \text{for all } p \in \mathcal{P}, a = (i,j) \in p \cap \mathcal{A}_{change} && (8.3) \\ y_i &\in \mathbb{N} \quad \text{for all } i \in \mathcal{E} \\ z_p &\in \{0,1\} \quad \text{for all } p \in \mathcal{P}. \end{aligned}$$

To justify that (TDM-A) is in fact a correct model for (TDM) we have to show that in all optimal solutions $z_p = 0$ if and only if all connections on path p are maintained. Before we do this, we present some properties of optimal solutions. The first is that a vehicle can only gain a delay, if there is some reason for this, i.e., if some source delay has occurred, or if the vehicle waits for some other delayed vehicle to maintain a connection.

Let a feasible solution of (TDM-A) be given. Let

$$\mathcal{A}^{fix} = \{a \in \mathcal{A} : \text{ there exists } p \in \mathcal{P} \text{ with } a \in p \text{ and } z_p = 0\}$$

denote the set of connections which are maintained in this solution. Then determine the time-minimal solution y^* with respect to $\mathcal{A}^{fix}$, i.e., an optimal solution of $(\text{TT}(\mathcal{A}^{fix}))$ according to page 110. We can be sure that the timetable of this solution is as tight as possible. Finally, we reconsider all paths to find out if some additional ones are maintained and adapt the solution of z accordingly. The solution obtained by this procedure is called a *reduced* solution of (TDM-A).

Definition 8.2. *Let (y, z) be a feasible solution of (TDM-A). Define*

$$y(z) = y(\mathcal{A}^{fix}(z)), \text{ where} \tag{8.4}$$
$$\mathcal{A}^{fix}(z) = \{a \in \mathcal{A} : \text{ there exists } p \in \mathcal{P} \text{ with } a \in p \text{ and } z_p = 0\} \tag{8.5}$$
$$z_p(y) = \begin{cases} 0 & \text{if } y_i - y_j \le s_a \text{ for all } a = (i,j) \in p \\ 1 & \text{otherwise} \end{cases} \tag{8.6}$$

and the **reduced solution**

$$R^A(y, z) = (y^{red}, z(y^{red})),$$

where $y^{red} = y(z)$ is the time-minimal solution with respect to $\mathcal{A}^{fix}(z)$.

The following properties hold.

Lemma 8.3. *Let (y, z) be a feasible solution of (TDM-A).*

1. *$(y(z), z)$ is feasible for (TDM-A) and $f_{\text{TDM-A}}(y(z), z) \le f_{\text{TDM-A}}(y, z)$.*
2. *Let $d_i \le D < T$ for all $i \in \mathcal{E}_{del}$. Then $R^A(y, z)$ is feasible for (TDM-A) and $f_{\text{TDM-A}}(R^A(y, z)) \le f_{\text{TDM-A}}(y, z)$.*

Proof. Let a feasible solution (y, z) be given.

1. Let us denote $y^{red} = y(z)$. Since y^{red} is feasible for $(\text{TT}(\mathcal{A}^{fix}))$ (see page 110) we obtain that (y^{red}, z) is feasible for (TDM-A): (8.1) holds due to (7.1), (8.2) and (8.3) are ensured due to (7.2) (take equation (8.5) in Definition 8.2 into account), and the integrality of the variables is also required in $\text{TT}(\mathcal{A}^{fix})$.
   Furthermore, we know that y is also feasible for $\text{TT}(\mathcal{A}^{fix})$ meaning that we can use part 1 of Lemma 7.8 (see page 114) and conclude that $y_i^{red} \le y_i$ for all $a \in \mathcal{E}$. Hence, $f_{\text{TDM-A}}(y^{red}, z) \le f_{\text{TDM-A}}(y, z)$.
2. Consider $R^A(y, z) = (y(z), z(y(z)))$. From part 1 of this lemma, we know that $(y(z), z) = (y^{red}, z)$ is feasible and the definition of $z^{red} = z(y^{red})$ gives us feasibility of $R(y, z) = (y^{red}, z(y^{red}))$. For the objective function, we first note that $z_p = 0$ means that $y_i^{red} - y_j^{red} \le s_a$ for all $a = (i, j) \in p$ is still satisfied for the time-reduced solution y^{red}, hence $z_p^{red} = z(y_p^{red}) = 0$. This means $z^{red} \le z$. Furthermore, $y_{i(p)} < T$ due to Corollary 7.9 and our assumption $d_i \le D < T$. We obtain:

$$\begin{aligned} f_{\text{TDM-A}}(y, z) &\ge f_{\text{TDM-A}}(y^{red}, z), \text{ see part 1 of this lemma,} \\ &\ge f_{\text{TDM-A}}(y^{red}, z(y^{red})) = f_{\text{TDM-A}}(R^A(y, z)) \\ &\quad \text{since } y^{red} < T \text{ and } z^{red} \le z. \end{aligned}$$

□

Note that $R^A(y, z) = R^A(R^A(y, z))$, i.e., reduced solutions cannot be further reduced. We now can justify the size of M in the formulation (TDM-A). Recall that $D = \max\{d_i : i \in \mathcal{E}\}$.

Lemma 8.4. *$M = D$ is large enough in (TDM-A).*

Proof. Let (y^*, z^*) be an optimal solution of (TDM-A). Due to Lemma 8.3 we can assume that $y^* = y(\mathcal{A}^{fix}(z^*))$ is time-minimal. Hence $y_i^* \le D$ (Corollary 7.9), and we obtain $y_i^* - y_j^* - s_a \le D$, yielding that $M \ge D$ is large enough. □

In the following, we always assume $M \ge D = \max\{d_i : i \in \mathcal{E}\}$

The given formulation of model (TDM-A) can be linearized (and weakened) by substituting the quadratic term $y_{i_p}(1 - z_p)$ by a new variable q_p, leading to the following model (TDM-B).

(TDM-B)

$$\min f_{\mathrm{TDM-B}} = \sum_{p \in \mathcal{P}} w_p(q_p + T z_p)$$

such that

$$\begin{aligned}
y_i &\ge d_i \quad \text{for all } i \in \mathcal{E}_{del} \\
y_i - y_j &\le s_a \quad \text{for all } a = (i,j) \in \mathcal{A}_{wait} \cup \mathcal{A}_{drive} \\
-M z_p + y_i - y_j &\le s_a \quad \text{for all } a = (i,j) \in \mathcal{A}_{change} \\
-M z_p + y_{i(p)} - q_p &\le 0 \quad \text{for all } p \in \mathcal{P} && (8.7)\\
q_p &\ge 0 \quad \text{for all } p \in \mathcal{P} && (8.8)\\
y_i &\in \mathbb{N} \quad \text{for all } i \in \mathcal{E} \\
z_p &\in \{0,1\} \quad \text{for all } p \in \mathcal{P}.
\end{aligned}$$

Lemma 8.5. *The linearization is correct.*

Proof. (TDM-A) $\Longrightarrow$ (TDM-B): Let (y, z) be a feasible solution of (TDM-A). According to Lemma 8.3 we can without loss of generality assume that (y, z) is a reduced solution. For all $p \in \mathcal{P}$ define $q_p = y_{i(p)}(1 - z_p)$. Since $y_{i(p)} \le D \le M$ we get for all $p \in \mathcal{P}$ that

$$-M z_p + y_{i(p)} \le -y_{i(p)} z_p + y_{i(p)} = q_p.$$

Hence, (y, z, q) is feasible for (TDM-B), and both solutions have the same objective value.

(TDM-B) $\Longrightarrow$ (TDM-A): Let (y, z, q) be a feasible solution of (TDM-B). Then (y, z) is also feasible for (TDM-A). From (8.7) and (8.8) we conclude that

$$\begin{aligned}
q_p &\ge y_{i_p} \text{ if } z_p = 0 \\
q_p &\ge 0 \quad \text{if } z_p = 1.
\end{aligned}$$

Consequently, $q_p \ge y_{i(p)}(1 - z_p)$, i.e., $f_{\mathrm{TDM-A}} \le f_{\mathrm{TDM-B}}$. □

8.2 Activity-based Model

Here we present another model for (TDM). By *activity-based* we mean that we do not focus on the paths $p \in \mathcal{P}$, but sum up the additional delay (i.e., the tension) on each single activity $a \in \mathcal{A}$. Therefore, we use variables $\bar{z}_a$ describing if the connection $a \in \mathcal{A}_{change}$ is missed ($\bar{z}_a = 1$) or maintained ($\bar{z}_a = 0$).

To motivate this new approach, consider some activity $a \in \mathcal{A} \setminus \mathcal{A}_{change}$. We want to calculate the additional delay customers will get while using this activity. The delay customers already have at the start of $a = (i,j)$ is y_i, and at the end of a the delay is y_j. This means, the tension $t_a = t_a^y = y_j - y_i$ is the additional delay gained by the customers while performing this activity. Note that t_a can be negative, meaning that slack times are used to compensate an already existing delay. For changing arcs we have to be more careful. Let $a = (i,j) \in \mathcal{A}_{change}$ and suppose first, that a is maintained. Then the additional delay on a is again given by the tension $t_a = t_a^y = y_j - y_i$. On the other hand, if a is missed, the additional delay for the customers who planned to use activity a is given by $T - y_i = t_a + T - y_j$, since they now have to wait the remaining time period until the next (non-delayed) vehicle arrives for carrying on their journey.

The main idea of the second model is to add up all these single delays over all activities. To this end, we have to extend the set of events and activities similar to Section 7.2, and we also extend the paths as follows: For each $p \in \mathcal{P}$, $p = (i_1^p, i_2^p, \ldots, i_L^p)$ add one common additional event s representing the arrival of the customers at their first station (by a means of transport which is not considered in the delay management problem). Furthermore, we add *timetable arcs*

$$a = (s, i_1^p)$$

for each path p. The resulting network resembles the CPM-network $\mathcal{N}^s = (\mathcal{E}^s, \mathcal{A}^s(\mathcal{A}_{change}))$ (see Notation 7.4 on page 112) in the special case that $\mathcal{A}^{fix} = \mathcal{A}_{change}$ with

$$\begin{aligned}
\mathcal{E}^s &= \mathcal{E} \cup \{s\} \\
\mathcal{A}^s &= \mathcal{A}^s(\mathcal{A}_{change}) = \mathcal{A} \cup \{(s,i) : i \in \mathcal{E}\} \text{ and} \\
\mathcal{P}^s &= \{(s, i_1^p, \ldots, i_L^p) : p \in \mathcal{P}\}.
\end{aligned}$$

This construction makes sure that the delay of a customer waiting at some station for his first (delayed) vehicle to come, is taken into account. We always assume that customers reach their first station without any delay, i.e.,

$$y_s = 0.$$

Now we can present the new model. As before, we assume that $T, M \geq D$. The following decision variables are necessary for (TDM-C).

$$y_i = \text{delay of event } i,$$

as before, and

$$\bar{z}_a = \begin{cases} 0 & \text{if connection } a \text{ is maintained} \\ 1 & \text{otherwise,} \end{cases}$$

$$\tilde{z}_a^p = \begin{cases} 1 & \text{if activity } a \text{ is reached on path } p \text{ without any missed} \\ & \text{connection before} \\ 0 & \text{otherwise,} \end{cases}$$

$$w_a = \text{number of customers who } \textit{really} \text{ use activity } a.$$

It is important to note that the number of customers w_a (really) using activity $a \in \mathcal{A}$ is a variable, since it depends on the wait-depart decisions whether customers using a path $p \in \mathcal{P}^s$ will reach all arcs $a \in p$ or not.

(TDM-C)

$$\min f_{\text{TDM-C}} = \sum_{a=(i,j)\in\mathcal{A}^s} w_a(y_j - y_i) + \sum_{a=(i,j)\in\mathcal{A}_{change}} w_a \bar{z}_a (T - y_j)$$

such that

$$y_i \geq d_i \quad \text{for all } i \in \mathcal{E}_{del} \tag{8.9}$$

$$y_i - y_j \leq s_a \quad \text{for all } a = (i,j) \in \mathcal{A}_{wait} \cup \mathcal{A}_{drive} \tag{8.10}$$

$$-M\bar{z}_a + y_i - y_j \leq s_a \quad \text{for all } a = (i,j) \in \mathcal{A}_{change} \tag{8.11}$$

$$\tilde{z}_a^p + \sum_{\substack{\tilde{a}\in p\cap\mathcal{A}_{change}: \\ \tilde{a}\prec a}} \bar{z}_{\tilde{a}} \geq 1 \quad \text{for all } p \in \mathcal{P}^s \text{ and } a \in p \tag{8.12}$$

$$\tilde{z}_a^p + \bar{z}_{\tilde{a}} \leq 1 \quad \text{for all } p \in \mathcal{P}^s \text{ and for all } a, \tilde{a} \in p \text{ with } \tilde{a} \in \mathcal{A}_{change} \text{ and } \tilde{a} \prec a \tag{8.13}$$

$$w_a = \sum_{p\in\mathcal{P}^s: a\in p} w_p \tilde{z}_a^p \quad \text{for all } a \in \mathcal{A}^s \tag{8.14}$$

$$y_i \in \mathbb{N} \quad \text{for all } i \in \mathcal{E}$$

$$\bar{z}_a \in \{0,1\} \quad \text{for all } a \in \mathcal{A}^s$$

$$\tilde{z}_a^p \in \{0,1\} \quad \text{for all } p \in \mathcal{P}^s, a \in \mathcal{A}^s$$

$$w_a \in \mathbb{N} \quad \text{for all } a \in \mathcal{A}^s.$$

In the objective function the additional amount of delay on each activity is multiplied by the number of customers *really* using it. Restrictions (8.9) and (8.10) are the same as the first two restrictions in (TDM-A). In restriction (8.11) we make sure that $\bar{z}_a = 1$ if the connection $a \in \mathcal{A}_{change}$ is missed. Restriction (8.12) defines the values of $\tilde{z}_a^p$ in such a way, that they are forced to be 1, if no connection on path p before a has been missed, and (8.13) makes sure that $\tilde{z}_a^p = 0$ for all activities a after a missed connection $\tilde{a}$ on path p. Finally, in (8.14) the number of customers really using activity a is calculated by summing up the number of customers really using a over all paths containing a. In the following we will show that this model is equivalent to (TDM-A) and (TDM-B). Note that in this model we can easily relax the assumption of one common time period for all vehicles, but can introduce different time periods T_a for all activities $a \in \mathcal{A}_{change}$. Nevertheless, we will use only one common period T in the following.

Relation to (TDM-A)

Before we discuss the connection between (TDM-A) and (TDM-C) we show how the $\bar{z}, \tilde{z}, w$-variables can be constructed from a **given** feasible solution $y_i, i \in \mathcal{E}$. Consider (TDM-C) for some given y.

(TDM-C(y))

$$\min \sum_{a=(i,j)\in\mathcal{A}^s} w_a(y_j - y_i) + \sum_{a=(i,j)\in\mathcal{A}_{change}} w_a(T - y_j)\bar{z}_a$$

such that

$$\begin{aligned}
-M\bar{z}_a + y_i - y_j &\le s_a \quad \text{for all } a=(i,j)\in\mathcal{A}_{change}\\
\tilde{z}_a^p + \sum_{\substack{\tilde{a}\in p\cap\mathcal{A}_{change}:\\ \tilde{a}\prec a}} \bar{z}_{\tilde{a}} &\ge 1 \quad \text{for all } p\in\mathcal{P}^s \text{ and } a\in p\\
\tilde{z}_a^p + \bar{z}_{\tilde{a}} &\le 1 \quad \text{for all } p\in\mathcal{P}^s \text{ and for all } a,\tilde{a}\in p\\
&\qquad \text{with } \tilde{a}\in\mathcal{A}_{change} \text{ and } \tilde{a}\prec a\\
w_a &= \sum_{p\in\mathcal{P}^s: a\in p} w_p \tilde{z}_a^p \quad \text{for all } a\in\mathcal{A}^s\\
\bar{z}_a &\in \{0,1\} \quad \text{for all } a\in\mathcal{A}_{change}\\
\tilde{z}_a^p &\in \{0,1\} \quad \text{for all } p\in\mathcal{P}^s, a\in\mathcal{A}^s\\
w_a &\in \mathbb{N} \quad \text{for all } a\in\mathcal{A}^s.
\end{aligned}$$

A feasible solution of (TDM-C(y)) can be constructed using the following rules.

Notation 8.6. *Let $y \in \mathbb{N}^{|\mathcal{E}|}$ satisfy constraints (8.9) and (8.10). Define the* **y-constructed** *solution $C(y)$ by*

$$C(y) = (\bar{z}^c, \tilde{z}^c, w^c),$$

where $\bar{z}^c = \bar{z}(y)$, $\tilde{z}^c = \tilde{z}(\bar{z}^c)$, and $w^c = w(\tilde{z}^c)$ are given by

$$\bar{z}_a(y) = \begin{cases} 0 & \text{if } y_i - y_j \le s_a \\ 1 & \text{otherwise} \end{cases} \quad \text{for all } a = (i,j) \in \mathcal{A}_{change}, \tag{8.15}$$

$$\tilde{z}_a^p(\bar{z}) = \max \left\{ 1 - \sum_{\substack{a \in p \cap \mathcal{A}_{change}: \\ \tilde{a} \prec a}} \bar{z}_{\tilde{a}}, 0 \right\} \quad \text{for all } p \in \mathcal{P}^s, a \in p, \tag{8.16}$$

$$w_a(\tilde{z}) = \sum_{p \in \mathcal{P}^s : a \in p} w_p \tilde{z}_a^p \quad \text{for all } a \in \mathcal{A}^s. \tag{8.17}$$

A solution constructed by the rules above is a feasible solution.

Lemma 8.7. *Let $y \in \mathbb{N}^{|\mathcal{E}|}$ be given, such that (8.9) and (8.10) hold. Then $(y, C(y))$ is feasible for (TDM-C).*

Proof. $\bar{z}$ is defined in such a way that (8.11) holds, and (8.14) is also satisfied. From (8.16) we know that

$$\tilde{z}_a^p \ge 1 - \sum_{\substack{a \in p \cap \mathcal{A}_{change}: \\ \tilde{a} \prec a}} \bar{z}_{\tilde{a}},$$

hence (8.12) holds. Finally, assume $a, \tilde{a} \in p$ with $\tilde{a} \in \mathcal{A}_{change}$, $\tilde{a} \prec a$ and $\bar{z}_{\tilde{a}} = 1$. Again from (8.16) we obtain that in this case $\tilde{z}_a^p = 0$, establishing (8.13). □

It also holds that this solution is optimal for (TDM-C(y)) if $y_i \le T$ for all $i \in \mathcal{E}$, but the proof for this will be provided later, see Corollary 8.12. First, we discuss the connection between (TDM-A) and (TDM-C).

Theorem 8.8. *Model (TDM-A) and Model (TDM-C) lead to the same set of optimal solutions $y \in \mathbb{R}^{|\mathcal{E}|}$.*

Proof. First, if

$$w_a = \sum_{p \in \mathcal{P}^s : a \in p} w_p \tilde{z}_a^p \quad \text{for all } a \in \mathcal{A}^s$$

the objective function of (TDM-C) can be reformulated to

$$\begin{aligned} f_{\text{TDM-C}} &= \sum_{a=(i,j) \in \mathcal{A}^s} w_a (y_j - y_i) + \sum_{a=(i,j) \in \mathcal{A}_{change}} w_a \bar{z}_a (T - y_j) \\ &= \sum_{a=(i,j) \in \mathcal{A}^s} \sum_{p \in \mathcal{P}^s : a \in p} w_p \tilde{z}_a^p (y_j - y_i) + \sum_{a=(i,j) \in \mathcal{A}_{change}} \sum_{p \in \mathcal{P}^s : a \in p} w_p \tilde{z}_a^p \bar{z}_a (T - \\ &= \sum_{p \in \mathcal{P}^s} w_p \left(\sum_{a=(i,j) \in \mathcal{A}^s : a \in p} \tilde{z}_a^p (y_j - y_i) + \sum_{\substack{a=(i,j) \in \mathcal{A}_{change} \\ a \in p}} \tilde{z}_a^p \bar{z}_a (T - y_j) \right) \\ &=: \sum_{p \in \mathcal{P}^s} w_p C_p. \end{aligned}$$

For the objective of (TDM-A), we define

$$A_p = y_{i(p)}(1 - z_p) + T z_p.$$

(TDM-C) $\Longrightarrow$ (TDM-A): Let $(y, \bar{z}, \tilde{z}, w)$ be feasible for (TDM-C). Define $z_p = z_p(\bar{z})$ as follows:

$$z_p(\bar{z}) = \begin{cases} 0 & \text{if } \bar{z}_a = 0 \text{ for all } a \in p \cap \mathcal{A}_{change} \\ 1 & \text{otherwise.} \end{cases} \tag{8.18}$$

Then (8.1) holds due to (8.9), (8.2) holds due to (8.10), and (8.3) is trivially satisfied, if $z_p = 1$, and for $z_p = 0$ we know that $\bar{z}_a = 0$ for all $a \in p$ and hence (8.3) holds because of (8.11). This means (y, z) is feasible for (TDM-A). It remains to show that $A_p \le C_p$. To this end, let $p = (s, i_1, \ldots, i_L) \in \mathcal{P}^s$ be a path with $i(p) = i_L$.

Case 1: $\bar{z}_a = 0$ for all $a \in p \cap \mathcal{A}_{change}$. Then, we defined $z_p = 0$. From (8.12) we get that $\tilde{z}_a^p = 1$ for all $a \in p$. Hence, from Lemma 7.11 and since $y_s = 0$ we conclude that

$$C_p = \sum_{a=(i,j)\in\mathcal{A}^s : a \in p} y_j - y_i = y_{i_L} - y_s = A_p.$$

Case 2: There exists $a \in p \cap \mathcal{A}_{change}$ with $\bar{z}_a = 1$. Choose a minimal with respect to $\prec$ with this property, say

$$\bar{a} - (i_{\bar{k}-1}, i_{\bar{k}}).$$

Then, since $\bar{z}_a, \tilde{z}_a^p$ satisfy (8.12) and (8.13) we obtain

$$\tilde{z}_a^p = 0 \text{ for all } a \in p \text{ with } \bar{a} \prec a$$
$$\tilde{z}_a^p = 1 \text{ for all } a \in p \text{ with } a \preceq \bar{a}.$$

Hence, for all $a \in \mathcal{A}_{change} \cap p$ we get

$$\tilde{z}_a^p \bar{z}_a = \begin{cases} 1 & \text{if } a = \bar{a} \\ 0 & \text{otherwise.} \end{cases}$$

This yields

$$\begin{aligned} C_p &= \sum_{\substack{a=(i,j)\in\mathcal{A}^s : a \in p \\ \text{and } a \preceq \bar{a}}} y_j - y_i + (T - y_{i_{\bar{k}}}) \\ &= y_{i_{\bar{k}}} - y_{i_0} + T - y_{i_{\bar{k}}} = T = A_p. \end{aligned}$$

Together,

$$f_{\text{TDM-C}}(y, \bar{z}, \tilde{z}, w) = f_{\text{TDM-A}}(y, z(\bar{z})). \tag{8.19}$$

(TDM-A) $\Longrightarrow$ (TDM-C): Now let a feasible solution (y, z) of (TDM-A) be given. We assume that $y_i \le T$ for all $i \in \mathcal{E}$, otherwise we can take a time-minimal solution instead of (y, z) with equal or better objective value according to Lemma 8.3 on page 123, and know that this new solution satisfies $y_i \le D \le T$, see Corollary 7.9 (page 114). Since y satisfies (8.1) and (8.2) we can apply Lemma 8.7 and derive some feasible solution for (TDM-C) according to (8.15),(8.16), and (8.17). For the objective values of this solution we again look at C_p and A_p for some path $p = (s, i_1, \ldots, i_L) \in \mathcal{P}^s$ and get:

Case 1: If $z_p = 0$, we get from (8.3) that $y_i - y_j \le s_a$ for all $a = (i, j) \in p$. Hence, due to the definition of $\bar{z}_a$ we conclude that $\bar{z}_a = 0$ for all $a \in p \cap \mathcal{A}_{change}$, yielding

$$C_p = y_{i(p)} = A_p$$

analogously to Case 1 of the first part of the proof.

Case 2: Now consider the case that $z_p = 1$.

Case 2a: $y_i - y_j \le s_a$ for all $a = (i, j) \in p$, yielding that $\bar{z}_a = 0$ for all $a \in p$ and hence

$$C_p = y_{i(p)} \le T = A_p.$$

Case 2b: There exists $a = (i, j) \in p$ such that $y_i - y_j > s_a$. This gives us $\bar{z}_a = 1$, and choose $\bar{a} = (i_{\bar{k}-1}, i_{\bar{k}})$ minimal with respect to $\prec$ with this property. Then, from the definition of $\bar{z}^p_a$ we get

$$\bar{z}^p_a = 0 \text{ for all } a \in p \text{ with } \bar{a} \prec a$$
$$\bar{z}^p_a = 1 \text{ for all } a \in p \text{ with } a \preceq \bar{a}$$

and finally, analogously to Case 2 of the first part of the proof,

$$C_p = T = A_p.$$

Together, these three cases give us that for $y_i \le T$, $i \in \mathcal{E}$:

$$f_{\mathrm{TDM-A}}(y, z) \ge f_{\mathrm{TDM-C}}(y, C(y)). \tag{8.20}$$

Putting both directions together yields that there exists an optimal solution for (TDM-A) with perturbed timetable y if and only if there exists an optimal solution for (TDM-C) with the same perturbed timetable y. □

Reduced Solutions

We are now able to show how to improve a given feasible solution of (TDM-C) to a so-called *reduced* feasible solution, analogously to the concept of a reduced solution on page 123. The motivation to deal with reduced solutions is to exclude such feasible solutions which are obviously not optimal. For example, if the y variables are obviously too large, or if $\bar{z}$ is not chosen optimally. To exclude the first problem, we can use the time-minimal solution instead of the given one. This has already been formalized in Chapter 7 by defining *time-minimal* solutions through (TT($\mathcal{A}^{fix}$)) (see page 110).

Definition 8.9. *Let $(y, \bar{z}, \tilde{z}, w)$ be a feasible solution of (TDM-C). Define the* **reduced** *solution*

$$R^C(y, \bar{z}, \tilde{z}, w) = (y^{red}, C(y^{red}))$$

by

$$\mathcal{A}^{fix}(\bar{z}) = \{a \in \mathcal{A}_{change} : \bar{z}_a = 0\} \tag{8.21}$$

$$y^{red} = y(\mathcal{A}^{fix}(\bar{z})), \tag{8.22}$$

and $C(y^{red})$ according to Notation 8.6 on page 127.

The following statement ensures that $R^C(R^C(y, \bar{z}, \tilde{z}, w)) = R^C(y, \bar{z}, \tilde{z}, w)$, i.e., a reduced solution cannot be further reduced.

Lemma 8.10. *Let $(y^{red}, \bar{z}^{red}, \tilde{z}^{red}, w^{red}) = R^C(y, \bar{z}, \tilde{z}, w)$ be a reduced solution. Then*

- $C(y^{red}) = (\bar{z}^{red}, \tilde{z}^{red}, w^{red})$
- $y^{red} = y(\mathcal{A}^{fix}(\bar{z}^{red}))$.

Proof. The first statement follows directly from Definition 8.9. For the second statement we note that $\bar{z}^{red} \leq \bar{z}$, meaning that

$$\mathcal{A}^{fix}(\bar{z}) \subseteq \mathcal{A}^{fix}(\bar{z}^{red})$$

and hence TT($\mathcal{A}^{fix}(\bar{z})$) is a relaxation of TT($\mathcal{A}^{fix}(\bar{z}^{red})$). But since the optimal solution y^{red} of TT($\mathcal{A}^{fix}(\bar{z})$) is feasible for the non-relaxed optimization problem TT($\mathcal{A}^{fix}(\bar{z}^{red})$) it is the optimal solution, i.e., $y^{red} = y(\mathcal{A}^{fix}(\bar{z}^{red}))$. □

We finally can prove that the reduced solutions of (TDM-C) are never worse than the original ones, similarly to Lemma 8.3.

Theorem 8.11. *Let $d_i \leq D < T$ for all $i \in \mathcal{E}_{del}$. Then for each feasible solution $(y, \bar{z}, \tilde{z}, w)$ of (TDM-C) $R^C(y, \bar{z}, \tilde{z}, w)$ is feasible for (TDM-C) and*

$$f_{\text{TDM-C}}(R^C(y, \bar{z}, \tilde{z}, w)) \leq f_{\text{TDM-C}}(y, \bar{z}, \tilde{z}, w).$$

Proof. Let $(y, \bar{z}, \tilde{z}, w)$ be any feasible solution of (TDM-C). Replacing y by $y^{red} = y(\mathcal{A}^{fix}(\bar{z}))$ is still a feasible solution, since (8.9) and (8.10) hold due to (7.1) and (7.2), where the latter also ensures (8.11). The remaining conditions only depend on $\bar{z}, \tilde{z}$, and w, and thus they are also satisfied.
For showing that the objective function value of (TDM-C) does not increase we first remark that due to the optimality of y^{red} (for TT($\mathcal{A}^{fix}(\bar{z})$)) we have

$$y_i \geq y_i^{red} \text{ for all } i \in \mathcal{E} \tag{8.23}$$

due to part 1 of Lemma 7.8 (see page 114). Now we use the equivalence between (TDM-C) and (TDM-A). From $\bar{z}$ we construct $z(\bar{z})$ according to (8.18) on page 129, and note that both $(y, z(\bar{z}))$ and $(y^{red}, z(\bar{z}))$ are feasible for (TDM-A) since $(y, \bar{z}, \tilde{z}, w)$ and $(y^{red}, \bar{z}, \tilde{z}, w)$ are both feasible for (TDM-C) (see the first direction in the proof of Theorem 8.8).

Hence,

$$\begin{aligned}
f_{\text{TDM-C}}(y, \bar{z}, \tilde{z}, w) &= f_{\text{TDM-A}}(y, z(\bar{z})) \\
&\quad \text{due to (8.19), see page 129 in the proof of Theorem 8.8} \\
&= \sum_{p \in \mathcal{P}} w_p (y_{i(p)}(1 - z_p(\bar{z})) + T z_p(\bar{z})) \\
&\geq \sum_{p \in \mathcal{P}} w_p (y_{i(p)}^{red}(1 - z_p(\bar{z})) + T z_p(\bar{z})), \text{ see (8.23)} \\
&= f_{\text{TDM-A}}(y^{red}, z(\bar{z})) \\
&\geq f_{\text{TDM-C}}(y^{red}, C(y^{red})) \\
&\quad \text{due to (8.20), see page 130 in the proof of Theorem 8.8} \\
&= f_{\text{TDM-C}}(R^C(y, \bar{z})).
\end{aligned}$$

□

Using the equivalence of (TDM-A) and (TDM-C) it is now easy to show that $C(y)$ is an optimal solution of (TDM-C(y)), if $y_i \leq T$ for all $i \in \mathcal{E}$. In particular, replacing $(\bar{z}, \tilde{z}, w)$ by $C(y)$ will always yield the same or a better objective function value for (TDM-C).

Corollary 8.12. *Let $y_i \leq T$ for all $i \in \mathcal{E}$. Then $C(y)$ is the optimal solution of (TDM-C(y)).*

Proof. From Lemma 8.7 we know that $(y, C(y))$ is feasible for (TDM-C). Similar to the proof of Theorem 8.11 we get for all feasible solutions $(\bar{z}, \tilde{z}, w)$ of (TDM-C(y)) that

$$\begin{aligned}
f_{\text{TDM-C}}(y, \bar{z}, \tilde{z}, w) &\geq f_{\text{TDM-A}}(y, z(\bar{z})) \\
&\quad \text{due to (8.19), see page 129 in the proof of Theorem 8.8} \\
&\geq f_{\text{TDM-C}}(y, C(y)) \\
&\quad \text{due to (8.20), see page 130 in the proof of Theorem 8.8,}
\end{aligned}$$

where the latter inequality relies on the assumption $y_i \leq T$. □

On a first glance, (TDM-C) does not seem to be useful for solving the delay management problem better than (TDM-A), since it is much larger:

- (TDM-A) can be linearized (see Lemma 8.5) while (TDM-C) is cubic.
- The number of variables in (TDM-A) is $O(|\mathcal{P}|+|\mathcal{E}|)$, but $O((|\mathcal{P}||\mathcal{A}|+|\mathcal{E}|+|\mathcal{A}_{change}|)$ in (TDM-C).

But note again that (TDM-C) is more general since it allows one to replace the common time period T by time periods T_a for each changing activity $a \in \mathcal{A}_{change}$. Even with a common time period T we will need (TDM-C) to derive and solve a special case of (TDM) in the next section. We conclude this section by using (TDM-C) to derive a reduction result for (TDM) based on the following idea: Assume that the slack times are so large that the delay disappears after a few activities. Then we need not consider events which can not gain any delay in the worst-case time-minimal solution.

Lemma 8.13. *Let $y = y(\mathcal{A}_{change})$ be an optimal solution of $TT(\mathcal{A}_{change})$. Then there exists an optimal solution $(y^*, \bar{z}^*, \tilde{z}^*, w^*)$ of (TDM-C) such that*

- *For all $i \in \mathcal{E}$: If $y_i = 0$ then $y_i^* = 0$.*
- *For all $a = (i, j) \in \mathcal{A}_{change}$: If $y_i = 0$ then $\bar{z}_a^* = 0$.*

Proof. Using Theorem 8.11 we can assume that $(y^*, \bar{z}^*, \tilde{z}^*, w^*)$ is a reduced solution, in particular, $y^* = y(\mathcal{A}^{fix}(\bar{z}^*))$. Since $\mathcal{A}^{fix}(\bar{z}^*) \subseteq \mathcal{A}_{change}$ we know from part 2 of Lemma 7.8 that $y^* \le y$, in particular, $y_i^* = 0$ if $y_i = 0$. Furthermore, $\bar{z}^*$ satisfies (8.15) (see page 128). Hence $\bar{z}_a^* = 0$ since $y_i^* - y_j^* \le 0 \le s_a$. □

This kind of reduction will be referred to as *late reduction.*

Notation 8.14. *$\mathcal{E}_{red-late} = \{i \in \mathcal{E} : y_i(\mathcal{A}_{change}) > 0\}$ denotes the set of events that need to be considered.*

In the following we always can assume that $\mathcal{E} = \mathcal{E}_{red-late}$. (Note that in real-world instances, late reduction often leads to significantly smaller networks.)

8.3 Constant Weights and the Never-meet Property

The simplification we suggest for solving (TDM-C) is to fix the weights w_a as parameters instead of calculating them during the optimization. Doing so, we obtain the *total delay management problem with constant weights* . Its formulation is given by deleting constraints (8.12), (8.13), and (8.14) in (TDM-C), and fixing

$$w_a = \sum_{p \in \mathcal{P}^s : a \in p} w_p \quad \text{for all } a \in \mathcal{A}^s$$

as parameters, i.e., setting w_a as the "traffic load" on activity a. We obtain the following program.

$$\min f_{\text{TDM-const}'} = \sum_{a=(i,j)\in\mathcal{A}^s} w_a(y_j - y_i) + \sum_{a=(i,j)\in\mathcal{A}_{change}} w_a \bar{z}_a (T - y_j)$$

such that

$$\begin{aligned} y_i &\geq d_i \quad \text{for all } i \in \mathcal{E}_{del} \\ y_i - y_j &\leq s_a \quad \text{for all } a = (i,j) \in \mathcal{A}_{wait} \cup \mathcal{A}_{drive} \\ -M\bar{z}_a + y_i - y_j &\leq s_a \quad \text{for all } a = (i,j) \in \mathcal{A}_{change} \\ y_i &\in \mathbb{N} \ \ \forall i \in \mathcal{E} \\ \bar{z}_a &\in \{0,1\} \ \ \forall a \in \mathcal{A}^s. \end{aligned}$$

Again, $M = \max_{i\in\mathcal{E}} d_i$ is large enough.

In the following we will identify cases where we make no mistake by fixing the weights as proposed above. In the general case, if we process the decisions iteratively we can adapt the weights according to the decisions that have been fixed in preceding steps. This idea will be utilized in the branch and bound approach (Algorithm 20) in Section 8.6. We can further rewrite $f_{\text{TDM-const}'}$ as follows. For $i \in \mathcal{E}$ let

$$w_i = \sum_{p\in\mathcal{P}:i(p)=i} w_p \tag{8.24}$$

be the number of customers with final destination i. For all changing activities $a \in \mathcal{A}_{change}$ we further use

$$\begin{aligned} w_a &= \sum_{p\in\mathcal{P}^s:a\in p} w_p \\ &= \sum_{p\in\mathcal{P}:a\in p} w_p \end{aligned} \tag{8.25}$$

as the number of customers who plan to use changing activity a. Since

$$\begin{aligned} \sum_{a=(i,j)\in\mathcal{A}^s} w_a(y_j - y_i) &= \sum_{p\in\mathcal{P}^s} w_p \sum_{a=(i,j)\in p} y_j - y_i \\ &= \sum_{p\in\mathcal{P}} w_p(y_{i(p)} - y_s) \\ &\qquad \text{due to Lemma 7.11, see page 115} \\ &= \sum_{i\in\mathcal{E}} \sum_{\substack{p\in\mathcal{P}: \\ i(p)=i}} w_p y_i \\ &= \sum_{i\in\mathcal{E}} w_i y_i \end{aligned}$$

we rewrite

$$f_{\mathrm{TDM-const}'} = \sum_{a=(i,j)\in\mathcal{A}^s} w_a(y_j - y_i) + \sum_{a=(i,j)\in\mathcal{A}_{change}} w_a \bar{z}_a (T - y_j)$$
$$= \sum_{i\in\mathcal{E}} w_i y_i + \sum_{a=(i,j)\in\mathcal{A}_{change}} w_a \bar{z}_a (T - y_j).$$

Unfortunately, in general, we make a mistake by fixing the weights as above. This is illustrated in Figure 8.2 as follows.

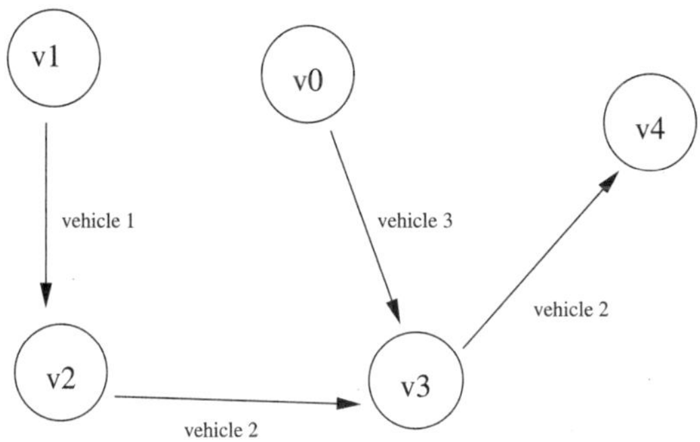

Fig. 8.2. A PTN in which (TDM-const) is not correct.

We assume that there are three vehicles 1,2, and 3, where vehicle 1 and vehicle 3 reach the stations v_2 and v_3 with a delay. We consider a path $p = (v_1, 1, v_2, 2, v_3, 2, v_4)$. Customers on this path use vehicle 1 until they reach station v_2; here they wish to change to vehicle 2 and to go on to stations v_3 and v_4. Suppose that vehicle 2 is not waiting for vehicle 1 at station v_2, such that the path p is missed. Assume further that vehicle 2 waits for the delayed vehicle 3 at station v_3. If we have not adapted the weights, the customers on path p are counted twice: First, since they missed their connection at station v_2, and secondly, since they reach their final destination v_4 with a delay. This double counting might influence decisions in the wrong way.

Fortunately, we are able to identify problem instances for which the model with constant weights is correct. One trivial case is, if no path in $\mathcal{P}$ contains a changing arc, i.e., no customer plans to change. A more interesting case, in which we make no mistake by using the constant weights will be described next.

Since $f_{\mathrm{TDM-const}'}$ still is no linear function we first further simplify the model. We would like to forget about subtracting y_j in the second part of the objective, to obtain a **linear** program. As usual, let $M \geq D$ and recall that

$$\text{for } i \in \mathcal{E}: \; w_i = \sum_{p \in \mathcal{P}: i(p)=i} w_p \text{ and}$$

$$\text{for } a \in \mathcal{A}_{change}: \; w_a = \sum_{p \in \mathcal{P}: a \in p} w_p$$

are fixed parameters in the following model.

(TDM-const)

$$\min f_{\text{TDM-const}} = \sum_{i \in \mathcal{E}} w_i y_i + \sum_{a \in \mathcal{A}_{change}} w_a T \bar{z}_a$$

such that

$$y_i \geq d_i \quad \text{for all } i \in \mathcal{E}_{del} \tag{8.26}$$

$$y_i - y_j \leq s_a \quad \text{for all } a = (i,j) \in \mathcal{A}_{wait} \cup \mathcal{A}_{drive} \tag{8.27}$$

$$-M\bar{z}_a + y_i - y_j \leq s_a \quad \text{for all } a = (i,j) \in \mathcal{A}_{change} \tag{8.28}$$

$$y_i \in \mathbb{N} \;\; \forall i \in \mathcal{E}$$

$$\bar{z}_a \in \{0,1\} \;\; \text{for all } a \in \mathcal{A}_{change}.$$

Surprisingly, (TDM-const) is equivalent to (TDM) in a large class of practical examples. Before we define this class of problems and prove the correctness of (TDM-const) in these cases (see Theorem 8.21), we show that each feasible solution of (TDM-const) yields an upper bound on (TDM).

Lemma 8.15. *Let $(y, \bar{z})$ be feasible for (TDM-const). Let f_{TDM} be the optimal objective value for (TDM). Then*

$$f_{\text{TDM-const}}(y, \bar{z}) \geq f_{\text{TDM}}.$$

Proof. Let $(y, \bar{z})$ be a feasible solution of (TDM-const). It can be extended to a feasible solution of (TDM-C) by defining $\bar{z}^c = \bar{z}(\bar{z})$ and $w^c = w(\bar{z})$ according to (8.16) and (8.17). Calculating also $z(\bar{z})$ finally gives us a feasible solution $(y, z(\bar{z}))$ of (TDM-A) and the *correct* weights w_a^c and w_i^c for the solution $(y, \bar{z})$, i.e.,

$$w_i^c = \sum_{p \in \mathcal{P}: i(p)=i} w_p z_p \text{ for all } i \in \mathcal{E} \tag{8.29}$$

$$w_a^c = \sum_{p \in \mathcal{P}: a \in p} w_p \bar{z}_a^p \text{ for all } a \in \mathcal{A}_{change}. \tag{8.30}$$

We know that the correct weights w^c are smaller as the parameters w in (TDM-const), i.e.,

$$w_a^c \leq w_a \text{ and } w_i^c \leq w_i. \tag{8.31}$$

Moreover, note that

$$z_p(\bar{z}) \leq \sum_{\substack{a \in \mathcal{A}_{change}: \\ a \in p}} \tilde{z}^p_a \bar{z}_a, \tag{8.32}$$

since, if $z_p = 1$ there exists $a \in p$ with $z_a = 1$. For a minimal (w.r.t. $\prec$) with this property, we additionally know that $\tilde{z}^p_a = 1$, hence $\bar{z}_a \tilde{z}^p_a = 1 \geq z_p$. This gives us

$$\begin{aligned}
f_{\text{TDM-const}}(y, \bar{z}) &= \sum_{i \in \mathcal{E}} w_i y_i + \sum_{a \in \mathcal{A}_{change}} w_a T \bar{z}_a \\
&\geq \sum_{i \in \mathcal{E}} w^c_i y_i + \sum_{a \in \mathcal{A}_{change}} w^c_a T \bar{z}_a, \quad \text{see (8.31)} \\
&= \sum_{i \in \mathcal{E}} \sum_{\substack{p \in \mathcal{P} \\ i(p)=i}} w_p (1 - z_p) y_i + \sum_{a \in \mathcal{A}_{change}} \sum_{p \in \mathcal{P}: a \in p} w_p \tilde{z}^p_a \bar{z}_a T \\
&= \sum_{p \in \mathcal{P}} w_p \left((1 - z_p) y_{i(p)} + \sum_{a \in \mathcal{A}_{change}: a \in p} \tilde{z}^p_a \bar{z}_a T \right) \\
&\geq \sum_{p \in \mathcal{P}} w_p \left((1 - z_p) y_{i(p)} + w_p z_p T \right), \quad \text{see (8.32)} \\
&= f_{\text{TDM-A}}(y, z) = f_{\text{TDM-C}}(y, \bar{z}, \tilde{z}, w^c)
\end{aligned}$$

i.e., each feasible solution of (TDM-const) gives an upper bound on (TDM). □

To specify problem instances for which (TDM-const) is correct, we first need some technical details.

Suppose that a set of connections $\mathcal{A}^{fix} \subseteq \mathcal{A}_{change}$ which have to be maintained is given, and let y^* be the time-minimal solution respecting these connections, where we have set all slack times to zero. Recall that

$$\mathcal{A}(\mathcal{A}^{fix}) = \mathcal{A}_{wait} \cup \mathcal{A}_{drive} \cup \mathcal{A}^{fix},$$

see Notation 7.2 on page 111. We want to show that in this solution the delay spreads out along all paths emanating at a delayed event. To this end we first need the following notation.

Notation 8.16. *Let $y^0(\mathcal{A}^{fix})$ denote the optimal solution of (TT($\mathcal{A}^{fix}$)), where the slack times are all set to 0.*

The following properties hold in the case of zero slack times.

Lemma 8.17. *Let $\mathcal{A}^{fix} \subseteq \mathcal{A}_{change}$ and $y^0 = y^0(\mathcal{A}^{fix})$. Then*

1. *$y^0 \geq y(\mathcal{A}^{fix})$.*
2. *Let p be any path consisting of activities $a \in \mathcal{A}(\mathcal{A}^{fix})$ and let $i \in p$. If $y^0_i > 0$ then $y^0_j > 0$ for all $j \in p$ with $i \prec j$.*

Proof. 1. Induction. Let $y^0 = y^0(\mathcal{A}^{fix})$ and $y = y(\mathcal{A}^{fix})$.
Start: Let i be a minimal element of $\mathcal{E}$. Then $y_i^0 = 0 = y_i$.
Conclusion: Let the induction hypothesis be true for all $j \prec i$. Then

$$\begin{aligned} y_i^0 &= \max_{a=(j,i)\in\mathcal{A}(\mathcal{A}^{fix})} y_j^0 \\ &\geq \max_{a=(j,i)\in\mathcal{A}(\mathcal{A}^{fix})} y_j^0 - s_a \\ &\geq \max_{a=(j,i)\in\mathcal{A}(\mathcal{A}^{fix})} y_j - s_a = y_i. \end{aligned}$$

2. Again, we use induction. Let $p = (i = i_0, i_1, \ldots, i_L)$.
Start: $y_i^0 > 0$ due to the assumption.
Conclusion: Take $i_l \in p$, $l \geq 1$. Let the induction hypothesis be true for all $k < l$. Then

$$\begin{aligned} y_{i_l}^0 &= \max_{a=(j,i_l)\in\mathcal{A}(\mathcal{A}^{fix})} y_j^0 \\ &\geq y_{l-1}^0 > 0. \end{aligned}$$

□

We are now in a position to introduce the never-meet-property.

Definition 8.18. *The delay management problem has the* **never-meet property** *if for each $\mathcal{A}^{fix} \subseteq \mathcal{A}_{change}$ the time-minimal solution $y = y^0(\mathcal{A}^{fix})$ satisfies the following two conditions for all $j \in \mathcal{E}_{red-late}$:*

1. If $(i_1, j), (i_2, j) \in \mathcal{A}^{fix} \cup \mathcal{A}_{wait} \cup \mathcal{A}_{drive}$, and $y_{i_2} > 0$ then $y_{i_1} = 0$.
2. If $(i_1, j) \in \mathcal{A}$, and $d_j > 0$ then $y_{i_1} = 0$.

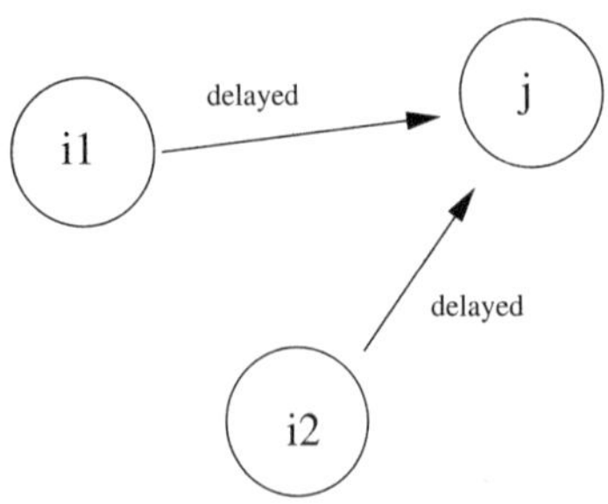

Fig. 8.3. Illustration of condition 1 of the never-meet property in $\mathcal{N}$.

Within the special structure of $\mathcal{N} = (\mathcal{E}, \mathcal{A})$ the first condition is always satisfied for $j \in \mathcal{E}_{arr}$, since each arrival event has exactly one incoming edge. For $j \in \mathcal{E}_{dep}$, at least one of the two edges (i_1, j) and (i_2, j) has to be in $\mathcal{A}_{change}$.

The situation that is **not** allowed to happen is depicted in Figure 8.3. The interpretation of the never-meet property is the following: By calculating the time-minimal solution with respect to some given $\bar{z}$, but without using slack-times, we can find out how far the source delays can spread out in this solution in the worst case. The never-meet property requires that in **no** feasible solution of (TDM) will two delayed vehicles meet, and that source delays can only occur after non-delayed events. Our goal is to show that we can fix the weights as in (TDM-const) without making any mistake, if the never-meet property holds.

Notation 8.19. *Let* $\mathcal{A}^{fix} \subseteq \mathcal{A}_{change}$ *and* $i \in \mathcal{E}$. *Define*

$$\mathcal{H}(i, \mathcal{A}^{fix}) = \{j \in \mathcal{E} : \text{ there exists a path from } i \text{ to } j \text{ consisting only of activities in } \mathcal{A}(\mathcal{A}^{fix})\} \cup \{i\}.$$

For all $j \in \mathcal{H}(i, \mathcal{A}^{fix})$ we hence have $i \preceq j$. An illustration of $\mathcal{H}(i, \mathcal{A}^{fix})$ is given in Figure 8.4.

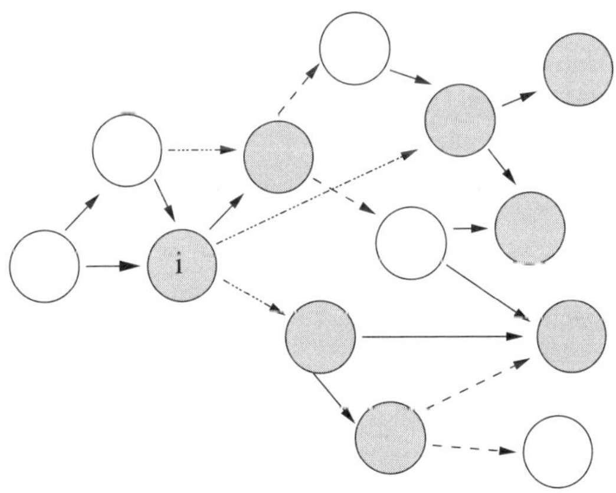

Fig. 8.4. Illustration of $\mathcal{H}(i, \mathcal{A}^{fix})$ (grey discs). The dashed arrows represent the non-fixed changing activities, the dotted arrows the fixed changing activities and the solid arrows waiting and driving activities.

We now can state the following property, which is important for proving the next theorem, but will also be needed for Algorithm 15 and for Algorithm 20.

Lemma 8.20. *Let (TDM) have the never-meet property and let* $(y, \bar{z}, \tilde{z}, w)$ *be a reduced feasible solution of (TDM-C). Let* $\tilde{a} = (\tilde{i}, \tilde{j}) \in \mathcal{A}_{change}$. *If* $\bar{z}_{\tilde{a}} = 1$ *we have the following.*

1. $y_i = 0$ *for all* $i \in \mathcal{H}(\tilde{j}, \mathcal{A}_{change})$,
2. $\bar{z}_a = 0$ *for all* $a = (i, j)$ *with* $i \in \mathcal{H}(\tilde{j}, \mathcal{A}_{change})$.

Proof. Let $\mathcal{H} = \mathcal{H}(\tilde{j}, \mathcal{A}_{change})$. First of all, note that

$$y_{\tilde{i}} - y_{\tilde{j}} > s_{\tilde{a}}$$

holds since $\bar{z}_a$ is given according to (8.15) and hence would have been set to 0 in the case that $y_{\tilde{i}} - y_{\tilde{j}} \leq s_{\tilde{a}}$. Especially we know that

$$y_{\tilde{i}} > 0. \tag{8.33}$$

Now add all changing activities with both endpoints in $\mathcal{H}$ to $\mathcal{A}^{fix}(\bar{z})$, i.e.,

$$\mathcal{A}^{fix}_{add} = \{a \in \mathcal{A}_{change} : \bar{z}_a = 0\} \cup \{a = (i,j) \in \mathcal{A}_{change} : i,j \in \mathcal{H}\}$$

and consider the solution $y^0 = y^0(\mathcal{A}^{fix}_{add})$ and the given reduced solution y, for which we know from Lemma 8.10 that

$$y = y(\mathcal{A}^{fix}(\bar{z})). \tag{8.34}$$

According to part 1 of Lemma 8.17 we obtain

$$y^0 \geq y(\mathcal{A}^{fix}_{add}) \geq y(\mathcal{A}^{fix}) = y,$$

where the second $\geq$ holds since $\mathcal{A}^{fix} \subseteq \mathcal{A}^{fix}_{add}$, see part 2 of Lemma 7.8. From (8.33) we hence get

$$y^0_{\tilde{i}} > 0. \tag{8.35}$$

Now consider $i \in \mathcal{H}$. We want to show that $y_i = 0$. Since $i \in \mathcal{H}$ there exists a path p from $\tilde{j}$ to i. In the solution y^0 without slack times, the delay $y_{\tilde{i}} > 0$ is transferred along p, since we added all connections on p to $\mathcal{A}^{fix}$, see part 2 of Lemma 8.17. We hence obtain

$$y^0_i > 0 \text{ for all } i \in \mathcal{H}. \tag{8.36}$$

From the never-meet property, and since $y^0_{\tilde{i}} > 0$ we hence conclude that

1. $y^0_j = 0$ for all $j \in \mathcal{E} \setminus \mathcal{H}$ such that there exists $i \in \mathcal{H}$ with $(j,i) \in \mathcal{A}$, and
2. $$d_i = 0 \text{ for all } i \in \mathcal{H}. \tag{8.37}$$

Consequently, since $y \leq y^0$ we obtain

$$y_j = 0 \quad \text{for all } j \in \mathcal{E} \setminus \mathcal{H} \text{ such that there exists } i \in \mathcal{H} \text{ with } (j,i) \in \mathcal{A}. \tag{8.38}$$

Again, recall that $y = y(\mathcal{A}^{fix}(\bar{z}))$ according to (8.34). Hence we can apply Corollary 7.6 (see page 113) and use that y satisfies

$$y_j = \max\{d_j, \max_{a=(i,j) \in \mathcal{A}(\mathcal{A}^{fix})} y_i - s_a\}.$$

To show that $y_i = 0$ for all $i \in \mathcal{H}$ we finally can use induction as follows.

Start: Take $\tilde{j}$ as minimal element of $\mathcal{H}$ (with respect to $\prec$). Then $\bar{z}_{\tilde{a}} = 1$ (due to the assumption in the lemma), such that $\tilde{a} \notin \mathcal{A}(\mathcal{A}^{fix})$. Furthermore, according to the never-meet property,
- $y_j = 0$ for all j with with $(j, \tilde{j}) \in \mathcal{A}(\mathcal{A}^{fix})$ (see (8.38)) and
- $d_{\tilde{j}} = 0$ (see (8.37)).

Hence,

$$y_{\tilde{j}} = \max\{d_{\tilde{j}}, \max_{a=(j,\tilde{j})\in\mathcal{A}(\mathcal{A}^{fix})} y_j - s_a\} = 0.$$

Conclusion: Consider $i \in \mathcal{H}$. Let $(j, i) \in \mathcal{A}(\mathcal{A}^{fix})$. Then, for $j \in \mathcal{H}$ the induction hypothesis yields $y_j = 0$ since this holds for all $j \in \mathcal{H}$ with $j \prec i$. On the other hand, for $j \notin \mathcal{H}$ we conclude $y_j = 0$ from (8.38). Furthermore, $d_i = 0$ (8.37). Together, we obtain

$$y_i = \max\{d_i, \max_{a=(j,i)\in\mathcal{A}(\mathcal{A}^{fix})} y_j - s_a\} = 0.$$

For the second condition of the lemma consider $a = (i, j) \in \mathcal{A}_{change}$ with $i \in \mathcal{H}$. Then $j \in \mathcal{H}$ and from part 1 of the lemma we conclude that $y_i = y_j = 0$, hence, $y_i - y_j \le s_a$ and $\bar{z}_a = 0$ since $\bar{z} = \bar{z}(y)$ according to (8.15). □

Theorem 8.21. *Model (TDM-const) is correct if the never-meet property holds.*

Proof. We show that (TDM-C) and (TDM-const) are equivalent in this case. First, given a solution $(y, \bar{z})$ of (TDM-const), we know from Lemma 8.15 that there exists a feasible solution of (TDM-C) with equal or better objective value.

The other direction is the interesting one: We show that each feasible solution of (TDM-C) corresponds to a feasible solution of (TDM-const) with the same or better objective value. More precisely, given some feasible solution of (TDM-C), let $(y, \bar{z}, \tilde{z}, w^c)$ denote the corresponding reduced feasible solution. We show that $y, \bar{z}$ is a feasible solution of (TDM-const) with the same objective value.
Feasibility of $y, \bar{z}$ for (TDM-const) is trivially satisfied. It remains to show that

$$f_{\mathrm{TDM-C}}(y, \bar{z}, \tilde{z}, w) = f_{\mathrm{TDM-const}}(y, \bar{z}).$$

To this end, suppose that for some $\bar{a} = (\bar{i}, \bar{j}) \in \mathcal{A}$

$$w_{\bar{a}} \neq w^{\mathrm{c}}_{\bar{a}}.$$

We have to show that in this case

$$y_{\bar{j}} - y_{\bar{i}} = 0,$$

and that for $\bar{a} \in \mathcal{A}_{change}$,

$$Tz_{\bar{a}} = 0,$$

meaning that the error we make by calculating the weights will not influence the value of the objective function. Note that this is satisfied if $\bar{i} \notin \mathcal{E}_{red-late}$, since we consider a time-minimal solution. From $w_{\bar{a}} \neq w^c_{\bar{a}}$ we get (by comparing (8.17) and (8.25), respectively), that

$$\sum_{p \in \mathcal{P}^s : \bar{a} \in p} w_p = w_{\bar{a}}$$
$$\neq w^c_{\bar{a}} = \sum_{p \in \mathcal{P}^s : \bar{a} \in p} w_p \tilde{z}^p_{\bar{a}}.$$

Hence there exists some path $p \in \mathcal{P}$ containing $\bar{a}$ such that $\tilde{z}^p_{\bar{a}} = 0$. Since we deal with a reduced solution we know that $\tilde{z}$ is given by (8.16) (on page 128), and hence there exists $\tilde{a} \in p$ with $\tilde{a} \prec \bar{a}$ and $\bar{z}_{\tilde{a}} = 1$. Without loss of generality let us take $\tilde{a} = (\tilde{i}, \tilde{j})$ minimal with this property, i.e., we choose the first changing activity on path p that is marked as missed. For an illustration, see Figure 8.5.

Fig. 8.5. The path p in the proof of Theorem 8.21. The grey events belong to $\mathcal{H}(\tilde{j}, \mathcal{A}_{change})$.

Since $\bar{i}, \bar{j} \in \mathcal{H}(\tilde{j}, \mathcal{A}_{change})$ we derive from Lemma 8.20 that

- $y_{\bar{i}} = y_{\bar{j}} = 0$, and
- if $\bar{a} \in \mathcal{A}_{change}$ then $\bar{z}_a = 0$.

Hence,

$$y_{\bar{j}} - y_{\bar{i}} = 0,$$

and if $\bar{a} \in \mathcal{A}_{change}$,

$$Tz_{\bar{a}} = 0,$$

which completes the proof. □

The next question is, how to find out efficiently, whether the never-meet property holds? To this end, we suggest the following algorithm.

Algorithm 13: Testing the never-meet property:

Input: $\mathcal{N}$, d_i, s_a.
Output: Answer *yes* or *no*.
Step 1. Calculate $y(\mathcal{A}_{change})$ by Algorithm 12. Let

$$\mathcal{E} = \mathcal{E} \cap \{i : y_i > 0\},$$
$$\mathcal{A} = \{(i,j) \in \mathcal{A} : i, j, \in \mathcal{E}\}).$$

Step 2. Calculate $y^0(\mathcal{A}_{change})$ for the reduced network $\mathcal{N} = (\mathcal{E}, \mathcal{A})$ by Algorithm 12.
Step 3. If the conditions of Definition 8.18 hold in this particular solution, the answer is *yes*, otherwise *no*.
Step 4. Stop.

The algorithm is based on the fact that it is enough to test the never-meet property in the worst case, namely, if all connections are maintained. If no two delayed vehicles will meet in this particular feasible solution, it can also not happen in any other feasible solution of (TDM) (with zero slack times). Formally, this is justified in the following theorem.

Theorem 8.22. *Algorithm 13 is correct.*

Proof. We have to show that conditions 1. and 2. of Definition 8.18 hold for all $\mathcal{A}^{fix} \subseteq \mathcal{A}_{change}$ if they hold for $\mathcal{A}^{fix} = \mathcal{A}_{change}$. Let $y^* = y^0(\mathcal{A}_{change})$ be the time-minimal solution of $\mathrm{TT}(\mathcal{A}_{change})$ with zero slack times, and let $y = y^0(\mathcal{A}^{fix})$ be the time-minimal solution of $\mathrm{TT}(\mathcal{A}^{fix})$ with zero slack times for any other $\mathcal{A}^{fix} \subseteq \mathcal{A}_{change}$. Then

$$y \le y^*,$$

according to part 2 of Lemma 7.8 (page 114). Consequently, for all $j \in \mathcal{E}$ we know the following.

1. If $(i_1, j), (i_2, j) \in \mathcal{A}$, and $y_{i_2} > 0$ then $y^*_{i_2} \ge y_{i_2} > 0$, hence $y^*_{i_1} = 0$, since y^* satisfies condition 1 of the never-meet property. But this means $0 \le y_{i_1} \le y^*_{i_1} = 0$.
2. If $(i_1, j) \in \mathcal{A}$, and $d_j > 0$ then $0 \le y_{i_1} \le y^*_{i_1} = 0$, again since y^* satisfies condition 2 of the never-meet property. □

So far, we have shown that (TDM-const) is correct, whenever the never-meet property holds, and that this property can be tested efficiently. Furthermore, our numerical results indicate that the never-meet property often is **almost** satisfied in practice. For example, a real-world situation of 120 delayed vehicles within the city of Kaiserslautern, each of them with a source delay of 10 minutes leads to only 148 conflicts with the never-meet property considering the set $\mathcal{U}_{10}$ as connections. A further analysis with randomly chosen source delays is shown in Figures 8.6 and 8.7. In both figures the graphed functions correspond to the different sets of relevant connections. The lowest function uses $\mathcal{U}_5$ as its set of connections, the next function corresponds to $\mathcal{U}_{10}$, then $\mathcal{U}_{30}$, and the top function refers to $\mathcal{U}_{60}$, confirming that the number of conflicts with the never-meet property increases if the set of connections is enlarged. Figure 8.6 shows the number of conflicts with the never-meet property as a function of the source delay, if we assume that 10 vehicles are delayed. It turns out that we can expect less than 50 conflicts if the source delays are smaller than 15 minutes.

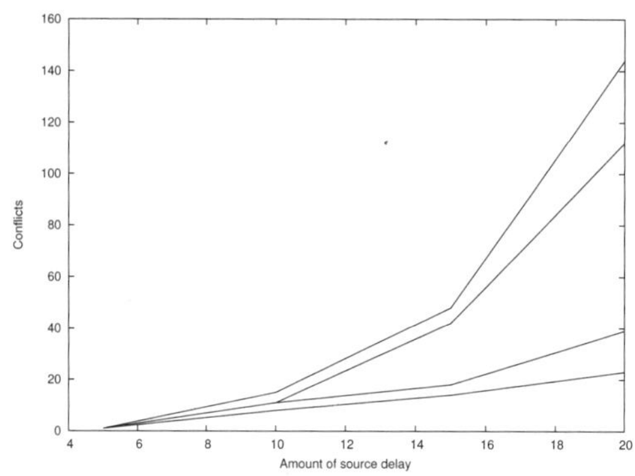

Fig. 8.6. Conflicts with the never-meet property as a function of the source delay, if 10 vehicles are delayed.

In Figure 8.7 the number of conflicts with the never-meet property is depicted as a function of the number of delayed vehicles. For this figure we assume a source delay of 15 minutes. Again, it turns out that not more than 50 conflicts are likely if the number of delayed vehicles is smaller than 10. The reason for the relatively low number of conflicts in practice is in particular due to the fact that only events in $\mathcal{E}_{red-late}$ that can gain a delay need to be considered (see Lemma 8.13 on page 133). Furthermore, most conflicts with the never-meet property arise at events within the city traffic included in our data, while the never-meet property is more likely to hold for transportation systems in a rural environment. A slightly more elegant formulation of the never-meet property and a detailed analysis of the never-meet property in a real-world railway example is presented in [Sch06].

But all this is only helpful if we can take advantage of the simplified model with constant weights in terms of solving it. In the next two sections we will

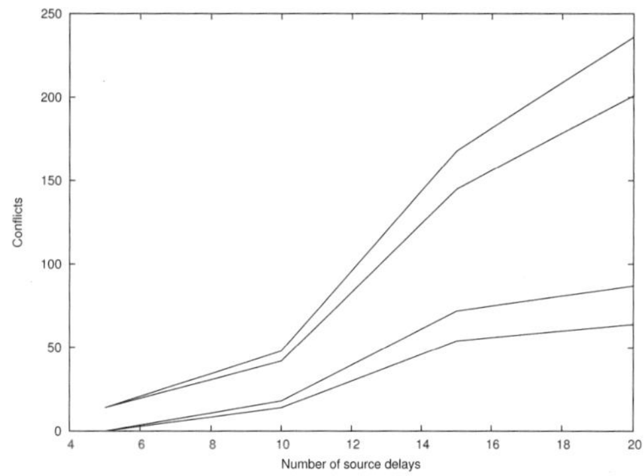

Fig. 8.7. Conflicts with the never-meet property as a function of the number of delayed vehicles, assuming a source delay of 15 minutes for each vehicle.

hence discuss (TDM-const) in more detail, while in Section 8.6 we will show how (TDM-const) can be used for solving the general problem (TDM).

8.4 A Simple Special Case

First, let us consider the special case (TDM-const-zero), in which

- all source delays have the same amount, i.e., $d_i \in \{0, D\}$ for all $i \in \mathcal{E}$, and
- all slack times are equal to zero, i.e., $s_a = 0$ for all $a \in \mathcal{A}$.

Let y be a time-minimal solution to this problem. Then, from Corollary 7.6 (see page 113) we conclude, that

$$y_i \in \{0, D\}$$

for all $i \in \mathcal{E}$. This means that we can use binary variables y_i instead of integer ones, with

$$y_i = \begin{cases} 1 & \text{if event } i \text{ is delayed by } D \\ 0 & \text{if event } i \text{ is not delayed.} \end{cases}$$

Consequently, $M = 1$ is large enough and (TDM-const) — even with the first objective $f_{\text{TDM-const}'}$ introduced on page 134 — simplifies to the following **linear** program. Recall that $\mathcal{E}_{del} = \{i \in \mathcal{E} : d_i > 0\}$ is the set of events with a source delay $d_i > 0$.

(TDM-const-zero)

$$\min \sum_{a=(i,j)\in\mathcal{A}^s} w_a D(y_j - y_i) + \sum_{a=(i,j)\in\mathcal{A}_{change}} w_a \bar{z}_a (T - D)$$

such that

$$-y_i \leq -1 \quad \text{for all } i \in \mathcal{E}_{del} \tag{8.39}$$

$$y_i - y_j \leq 0 \quad \text{for all } a = (i,j) \in \mathcal{A}_{wait} \cup \mathcal{A}_{drive} \tag{8.40}$$

$$\bar{z}_a + y_i - y_j \leq 0 \quad \text{for all } a = (i,j) \in \mathcal{A}_{change} \tag{8.41}$$

$$y_i \in \{0,1\} \ \ \forall i \in \mathcal{E}$$

$$\bar{z}_a \in \{0,1\} \ \ \forall a \in \mathcal{A}_{change},$$

where $w_a = \sum_{p \in \mathcal{P}^s : a \in p} w_p$ for all $a \in \mathcal{A}^s$ are given parameters as before (see, e.g., (8.25)).

Theorem 8.23. *(TDM-const-zero) can be solved in polynomial time.*

Proof. Let $C = |\mathcal{A}_{change}|$, $\bar{C} = |\mathcal{A}_{drive} \cup \mathcal{A}_{wait}|$ and $\bar{D} = |\mathcal{E}_{del}|$. Moreover, let I_K denote the unit matrix of size $K \times K$ and $O_{K,L}$ the zero matrix of size $K \times L$. Then the coefficient matrix of (TDM-const-zero) is

$$\Phi = \left(\begin{array}{c|c} -I_{\bar{D}} & 0_{\bar{D},C} \\ \hline \Theta^T & 0_{\bar{C},C} \\ & I_C \end{array} \right),$$

where the $|\mathcal{A}| \times |\mathcal{E}|$-matrix Θ^T is the transpose of the node-arc-incidence matrix Θ of $\mathcal{N}$ (as defined on page 115). Since Θ is totally unimodular (see [NW88], or Appendix A), Θ^T is. Adding a unit matrix on the right-hand side or above a totally unimodular matrix still yields a totally unimodular matrix (again, see Appendix A). Hence, Φ is totally unimodular. This means that all basic solutions of (TDM-const-zero) are integer and hence the LP-relaxation of (TDM-const-zero) can be used to solve the problem in polynomial time by linear programming methods. □

Note that (TDM-const-zero) is equivalent to the models developed independently in diploma theses by Kliewer [Kli00a] and Scholl [Sch01b], where the latter author also recognized the total unimodularity of the model.

The consequence of this result is that minimizing the total delay can be done efficiently if the never-meet property holds, all source delays are of the same amount, and all slack times are zero. If the slack times are not all zero, we can at least use (TDM-const-zero) as a bound.

Lemma 8.24. *Let $f_{\text{TDM-const-zero}}$ be the optimal objective value of (TDM-const-zero), and $f_{\text{TDM-const}}$ be the optimal objective value of (TDM-const). Then*

$$f_{\text{TDM-const-zero}} \geq f_{\text{TDM-const}}.$$

Proof. Let $(y, \bar{z})$ be any feasible solution of (TDM-const-zero). Defining $\tilde{y}_i = D$ for all $i \in \mathcal{E}$ with $y_i = 1$ yields a feasible solution $(\tilde{y}, \bar{z})$ for (TDM-const) with

$$f_{\text{TDM-const-zero}}(y, \bar{z}) = f_{\text{TDM-const}}(\tilde{y}, \bar{z}),$$

i.e., $f_{\text{TDM-const-zero}}(y, \bar{z})$ is an upper bound on (TDM-const) for all feasible solutions of (TDM-const-zero). □

8.5 Solving the model with constant weights

In this section we discuss the more general case of (TDM-const) as given on page 136, where we do not have zero slack times. Note that (TDM-const) (see page 136) without constraints (8.28) is efficiently solvable by the methods discussed in Chapter 7. Nevertheless, the Lagrange-relaxation of these constraints will not give any better bound than the LP-relaxation, since all extreme points of the relaxed feasible set

$$\{y \in \mathbb{R}^{|\mathcal{E}|} : y_i \geq d_i \text{ for all } i \in \mathcal{E} \text{ and}$$
$$y_i - y_j \leq s_a \text{ for all } a = (i,j) \in \mathcal{A}_{wait} \cup \mathcal{A}_{drive}\}$$

are integral according to Theorem 7.1, see integer programming textbooks such as [Wol98] or [NW88]. A brief review of this result is given in Appendix A. Accordingly, we investigate the LP-relaxation of (TDM-const), given by the following formulation.

(LP-TDM-const)

$$\min f_{\text{TDM-const}} = \sum_{i \in \mathcal{E}} w_i y_i + \sum_{a \in \mathcal{A}_{change}} w_a T \bar{z}_a$$

such that

$$y_i \geq d_i \quad \text{for all } i \in \mathcal{E}_{del} \tag{8.42}$$
$$y_i - y_j \leq s_a \quad \text{for all } a = (i,j) \in \mathcal{A}_{wait} \cup \mathcal{A}_{drive} \tag{8.43}$$
$$-M\bar{z}_a + y_i - y_j \leq s_a \quad \text{for all } a = (i,j) \in \mathcal{A}_{change} \tag{8.44}$$
$$y_i \geq 0 \ \ \forall i \in \mathcal{E}$$
$$\bar{z}_a \geq 0 \ \ \forall a \in \mathcal{A}^s.$$

We denote (as before) $C = |\mathcal{A}_{change}|$, $\bar{C} = |\mathcal{A}_{drive} \cup \mathcal{A}_{wait}|$, I_K as the unit matrix of size $K \times K$ and $O_{K,L}$ as the zero matrix of size $K \times L$. Then the coefficient matrix of (LP-TDM-const) is

$$\left(\begin{array}{c|c} 0_{|\mathcal{E}|+\bar{C},C} & \begin{array}{c} -I_{|\mathcal{E}|} \\ \hline -\Theta^T \end{array} \\ \hline -MI_C & \end{array}\right).$$

The transpose of the coefficient matrix is hence given by

$$\left(\begin{array}{c|c|c} O_{C,|\mathcal{E}|} & O_{C,\bar{C}} & -MI_C \\ \hline -I_{|\mathcal{E}|} & \multicolumn{2}{c}{-\Theta} \end{array}\right).$$

Since

$$\min\{cx : Ax \leq b, x \geq 0\} \quad \text{and} \quad \max\{-ub : -A^T u \leq c, u \geq 0\}$$

are dual to each other, we obtain the following dual program:

$$\max \sum_{i \in \mathcal{E}} d_i \xi_i - \sum_{a \in \mathcal{A}} s_a \eta_a$$

such that

$$\begin{aligned} M\eta_a &\leq w_a T \text{ for all } a \in \mathcal{A}_{change} \\ \xi_i - \sum_{a=(i,j)} \eta_a + \sum_{a=(j,i)} \eta_a &\leq w_i \text{ for all } i \in \mathcal{E} \\ \eta_a &\geq 0 \text{ for all } a \in \mathcal{A} \\ \xi_i &\geq 0 \text{ for all } i \in \mathcal{E}. \end{aligned}$$

The following observations about this dual of the relaxation can be made.

- For $i \notin \mathcal{E}_{del}$ we have that $d_i = 0$ and hence ξ_i has no contribution to the objective function. The best choice for ξ_i is hence $\xi_i = 0$.
- For $i \in \mathcal{E}_{del}$ we know that $y_i \geq d_i > 0$. The complementary slackness conditions hence yield that in the dual program

  $$\xi_i - \sum_{a=(i,j)} \eta_a + \sum_{a=(j,i)} \eta_a = w_i$$

  holds.

This means that the above formulation is equivalent to the following linear program.

(Dual-TDM-const)

$$\max \sum_{i \in \mathcal{E}_{del}} d_i \xi_i - \sum_{a \in \mathcal{A}} s_a \eta_a$$

such that

$$\begin{aligned} \eta_a &\leq \frac{w_a T}{M} \text{ for all } a \in \mathcal{A}_{change} \\ -\sum_{a=(i,j)} \eta_a + \sum_{a=(j,i)} \eta_a &\leq w_i \text{ for all } i \in \mathcal{E} \setminus \mathcal{E}_{del} \\ -\sum_{a=(i,j)} \eta_a + \sum_{a=(j,i)} \eta_a &= w_i - \xi_i \text{ for all } i \in \mathcal{E}_{del} \\ \eta_a &\geq 0 \text{ for all } a \in \mathcal{A} \\ \xi_i &\geq 0 \text{ for all } i \in \mathcal{E}. \end{aligned}$$

A feasible solution of (Dual-TDM-const) is given by $\xi_i = w_i$ for all $i \in \mathcal{E}_{del}$ and $\eta_a = 0$ for all $a \in \mathcal{A}$ with objective value $\sum_{i\in\mathcal{E}_{del}} d_i w_i$, yielding a lower bound on (LP-TDM-const) and hence on (TDM-const).

Lemma 8.25. *Let (η, ξ) be an optimal solution of (Dual-TDM-const), and $(y, \bar{z})$ be an optimal solution of (LP-TDM-const). Then*

1. $\xi_i \geq w_i$ *for all* $i \in \mathcal{E}_{del}$.
2. *If* $\bar{z}_a > 0$ *then* $\bar{z}_a = \frac{y_i - y_j - s_a}{M}$.

Proof. 1. Suppose $\xi_i < w_i$ for $i \in \mathcal{E}_{del}$. Increasing ξ_i to w_i for all $i \in \mathcal{E}_{del}$ implies that $\eta_a = 0$ for all $a \in \mathcal{A}$ is feasible. Since $d_i > 0$ for all $i \in \mathcal{E}_{del}$ the objective value has strictly increased, a contradiction.
2. Let $\bar{z}_a > 0$ for some $a \in \mathcal{A}_{change}$. Complementary slackness then requires that $\eta_a = \frac{w_a T}{M}$. Since $w_a > 0$ this yields $\eta_a > 0$ and hence, again from complementary slackness, we conclude that (8.44) has to be satisfied with equality, yielding $\bar{z}_a = \frac{y_i - y_j - s_a}{M}$. □

Note that (Dual-TDM-const) has the following interpretation as a flow problem in $\mathcal{N}$. First, assume the ξ_i variables as given. Then the η_a variables are flow variables for all activities $a \in \mathcal{A}$ with lower bound zero, and upper bounds $\frac{w_a T}{M}$ on changing arcs $a \in \mathcal{A}_{change}$. The goal is to minimize the costs of the flow, where the costs on the activities are given by s_a for all $a \in \mathcal{A}$. The status of the nodes $i \in \mathcal{E}_{del}$ depends on the value for ξ_i.

- If $\xi_i > w_i$ then i is a surplus node,
- if $\xi_i = w_i$, then i is a transshipment node, and
- i is a demand node if $\xi_i < w_i$.

From Lemma 8.25 we know that in an optimal solution, none of the $i \in \mathcal{E}_{del}$ is a demand node.

On the other hand, for $i \in \mathcal{E} \setminus \mathcal{E}_{del}$ node i can be a demand or a transshipment node, meaning that it can absorb demand until an amount of w_i but need not. Since the goal is to minimize the costs of the flow, it makes sense to leave as much flow in i as possible. This means, that, if the flow f_i into node $i \notin \mathcal{E}_{del}$ is known (i.e. $f_i = \sum_{a=(j,i)} \eta_i$), then the optimal choice of the η_a variables for all $a = (i, j)$ leaving node i satisfies

$$\sum_{a=(i,j)} \eta_a = \max\{0, f_i - w_i\},$$

if $\xi_i > w_i$ for all $i \in \mathcal{E}$, i.e., no demand has to be satisfied later on.

If the ξ_i variables are not given in advance, we add a source s and arcs (s, i) for all $i \in \mathcal{E}_{del}$. Then we interpret the variables ξ_i as flow variables on (s, i), and obtain that all nodes $i \in \mathcal{E}_{del}$ need to be transshipment nodes. The goal hence is to minimize the costs of the flow, but at the same time maximize the surplus of the sink, i.e. put as much flow into the network as possible.

Solving the relaxation provides a lower bound on (TDM-const) for which we can estimate its quality as follows.

Theorem 8.26. *Let $(y, \bar{z})$ be an optimal solution of the relaxation with objective value $f_{\text{LP-TDM-const}}$. Let f^* denote the optimal objective value of (TDM-const). Then*

$$0 \leq f^* - f_{\text{LP-TDM-const}} \leq T \sum_{a:\bar{z}_a>0} w_a \left(1 - \frac{y_i - y_j - s_a}{M}\right).$$

Proof. Let $\mathcal{A}^{fix} = \mathcal{A}^{fix}(\bar{z}) = \{a \in \mathcal{A}_{change} : \bar{z}_a = 0\}$ be the set of connections which are maintained in the optimal solution of the relaxation. From part 2 of Lemma 8.25 we know that

$$\bar{z}_a = \frac{y_i - y_j - s_a}{M} \text{ for all } a \notin \mathcal{A}^{fix}.$$

Since $M \geq D \geq y_i - y_j - s_a$ according to Corollary 7.9 we conclude that

$$\bar{z}_a \leq 1. \tag{8.45}$$

Furthermore, we obtain

$$\begin{aligned} f_{\text{LP-TDM-const}} &= \sum_{i \in \mathcal{E}} w_i y_i + \sum_{a \in \mathcal{A}_{change} \setminus \mathcal{A}^{fix}} T w_a \frac{y_i - y_j - s_a}{M} \\ &= \sum_{i \in \mathcal{E}} w_i y_i + \sum_{a:\bar{z}_a>0} T w_a \frac{y_i - y_j - s_a}{M}. \end{aligned}$$

We now define a feasible solution of (TDM-const) as

$$\begin{aligned} \bar{z}^u_a &= \begin{cases} 0 & \text{if } a \in \mathcal{A}^{fix} \\ 1 & \text{if } a \notin \mathcal{A}^{fix} \end{cases}, \text{ and} \\ y^u &= y(\mathcal{A}^{fix}). \end{aligned}$$

To compare the objective values of the two solutions $(y, \bar{z})$ and $(y^u, \bar{z}^u)$ we would like to replace y^u by y, but this is in general not feasible, since y need not be integer. Nevertheless, since $\bar{z}^u_a \geq \bar{z}_a$ (due to (8.45)) we have

$$-M\bar{z}^u_a + y_i - y_j \leq -M\bar{z}_a + y_i - y_j \leq s_a$$

for all $a = (i, j) \in \mathcal{A}_{change}$, such that $(y, \bar{z}^u)$ satisfies all constraints of (TDM-const), except for the integrality of y. Hence y is also a feasible solution of $\text{TT}(\mathcal{A}^{fix})$ (with $w^{fix}_i = w_i$, page 110). Consequently, $f_{\text{TT}(\mathcal{A}^{fix})}(y^u) \leq f_{\text{TT}(\mathcal{A}^{fix})}(y)$, i.e.

$$\sum_{i \in \mathcal{E}} w_i y^u_i \leq \sum_{i \in \mathcal{E}} w_i y_i.$$

Now let f^u denote the objective value of $(y^u, \bar{z}^u)$, i.e. $f^* \leq f^u$. We get:

$$\begin{aligned}
0 \leq f^* - f_{\text{LP-TDM-const}} &\leq f^u - f_{\text{LP-TDM-const}} \\
&= \sum_{i \in \mathcal{E}} w_i y_i^u + \sum_{a:\bar{z}_a > 0} w_a T - \sum_{i \in \mathcal{E}} w_i y_i - \sum_{a:\bar{z}_a > 0} w_a T \frac{y_i - y_j - s_a}{M} \\
&\leq T \sum_{a:\bar{z}_a > 0} w_a \left(1 - \frac{y_i - y_j - s_a}{M}\right).
\end{aligned}$$

□

The solution $(y^u, \bar{z}^u)$ which we have constructed in the proof will be used as an upper bound in the branch and bound approach, see Algorithm 14.

We remark that Theorem 8.26 provides an alternative proof of Theorem 8.23 (page 146), since for (TDM-const-zero) we obtain that a connection $a = (i, j)$ is missed if and only if $y_i = 1$ and $y_j = 0$. I.e., for a missed connection we obtain

$$\frac{y_i - y_j - s_a}{M} = \frac{y_i - y_j - 0}{1} = 1.$$

Consequently, we get

$$f^* - f_{\text{LP-TDM-const-zero}} \leq T \sum_{a:\bar{z}_a > 0} w_a \left(1 - \frac{y_i - y_j - s_a}{M}\right) = 0,$$

meaning that the relaxation solves (TDM-const-zero).

Looking at the bound we further see that the relaxation and the bound both get better if we use a smaller M. Note that $M' < M$ is still large enough if there exists an optimal solution (for the problem with large M) satisfying

$$M' \geq y_i - y_j - s_a$$

for all $a \in \mathcal{A}_{change}$. To strengthen the relaxation we hence replace M by smaller values M_a for all $a \in \mathcal{A}_{change}$. This gives the following formulation.

(TDM-const-strong)

$$\min f_{\text{TDM-const}} = \sum_{i \in \mathcal{E}} w_i y_i + \sum_{a \in \mathcal{A}_{change}} w_a T \bar{z}_a$$

such that

$$\begin{aligned}
y_i &\geq d_i \quad \text{for all } i \in \mathcal{E}_{del} \\
y_i - y_j &\leq s_a \quad \text{for all } a = (i,j) \in \mathcal{A}_{wait} \cup \mathcal{A}_{drive} \\
-M_a \bar{z}_a + y_i - y_j &\leq s_a \quad \text{for all } a = (i,j) \in \mathcal{A}_{change} \qquad (8.46) \\
y_i &\in \mathbb{N} \ \ \forall i \in \mathcal{E} \\
\bar{z}_a &\in \{0,1\} \ \ \forall a \in \mathcal{A}_{change}.
\end{aligned}$$

Lemma 8.27. *Let $y = y(\mathcal{A}_{change})$ and let $M \geq M_a \geq y_i - s_a$ for all $a = (i,j) \in \mathcal{A}_{change}$. Then (TDM-const-strong) and (TDM-const) are equivalent.*

Proof. It is clear that the feasible set of (TDM-const-strong) is contained in (TDM-const), such that each feasible solution of (TDM-const-strong) is also a feasible solution of (TDM-const) with the same objective value. On the other hand, take an optimal solution $(y, \bar{z})$ of (TDM-const). Due to Lemma 8.10 we can assume that y is a time-minimal solution, i.e., $y = y(\mathcal{A}^{fix}(\bar{z}))$. Note that $\mathcal{A}^{fix}(\bar{z}) \subseteq \mathcal{A}_{change}$, hence we know from part 2 of Lemma 7.8 that

$$y = y(\mathcal{A}^{fix}(\bar{z})) \leq y(\mathcal{A}_{change}).$$

This gives us that for all $a \in \mathcal{A}_{change}$,

$$\begin{aligned} M_a &\geq y_i(\mathcal{A}_{change}) - s_a \\ &\geq y_i - y_j - s_a, \end{aligned}$$

hence $(y, \bar{z})$ is feasible for (TDM-const-strong) with the same objective function value and the result follows. □

Now Theorem 8.26 can be extended easily to the following result, in which we use the relaxation of (TDM-const-strong) to derive a sharper bound.

Corollary 8.28. *Let $M_a \geq y_i(\mathcal{A}_{change}) - s_a$ for all $a \in \mathcal{A}_{change}$. Let $(y, \bar{z})$ be an optimal solution of the relaxation of (TDM-const-strong) with objective value $f_{\mathrm{LP-TDM-const-strong}}$. Let f^* denote the optimal objective value of (TDM-const). Then*

$$0 \leq f^* - f_{\mathrm{LP-TDM-const}} \leq T \sum_{a:\bar{z}_a > 0} w_a \left(1 - \frac{y_i - y_j - s_a}{M_a}\right).$$

Unfortunately, even in the case of the never-meet property and for the strong formulation (TDM-const-strong) the LP-relaxation need not find an integer solution, as the following example demonstrates. Let a network with three events $\mathcal{E} = \{1, 2, 3\}$ and one changing activity $a = (1, 2)$ be given as depicted in Figure 8.8.

- Let $d_1 = 15$ be the source delay of event 1, and
- assume $T = 30$.
- Furthermore, we consider the following weights: $w_{12} = w_a = 7$, $w_1 = 0$, $w_2 = 10$, and $w_3 = 12$.
- The slack times are given as $s_a = s_{12} = 5$ and $s_{23} = 1$.

Calculating $y = y(\mathcal{A}_{change})$ gives $y_1 = 15, y_2 = 10, y_3 = 9$ and hence

$$M_a = 15 - 5 = 10.$$

The LP-relaxation hence is

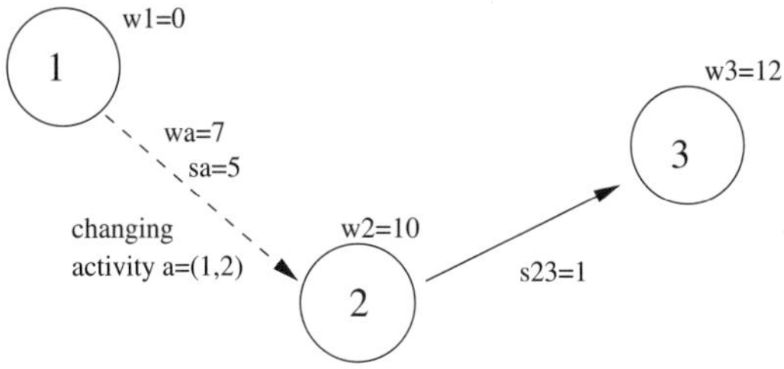

Fig. 8.8. The relaxation does not yield an integer solution.

$$\begin{aligned}\min 10y_2 + 12y_3 + 210\bar{z}_a & \\ \text{s.t. } y_1 &\geq 15 \\ -10\bar{z}_a + y_1 - y_2 &\leq 5 \\ y_2 - y_3 &\leq 1 \\ y_i, \bar{z}_a &\geq 0.\end{aligned}$$

Now compare the following three feasible (time-minimal) solutions

- $\bar{z}_a = 1, y = y(\bar{z})$, i.e., $y_1 = 15$, $y_2 = 0$, $y_3 = 0$ with objective value $f = 210$,
- $\bar{z}_a = 0, y = y(z)$, i.e., $y_1 = 15$, $y_2 = 10$, $y_3 = 9$ with objective value $f = 208$,
- $\bar{z}_a = \frac{1}{5}$, $y_1 = 15$, $y_2 = 8$, $y_3 = 7$ with objective value $f = 42+80+84 = 206$,

showing that the solution with $\bar{z} \notin \{0,1\}$ yields a better objective function value than all possible solutions with integer $\bar{z}_a$. Consequently, the optimal solution of the relaxation is not integer.

For the following branch and bound approach we not only need lower bounds, but also upper bounds. These are given by fixing some of the components of $\bar{z}$ in any node of the branch and bound tree. A feasible solution can then be calculated by solving TT($\mathcal{A}^{fix}(\bar{z})$). We consequently obtain $y = y(\mathcal{A}^{fix}(\bar{z}))$ and z according to (8.6) on page 123, i.e., by setting

$$z_p = \begin{cases} 0 & \text{if } y_i - y_j \leq s_a \text{ for all } a = (i,j) \subset p \\ 1 & \text{otherwise.} \end{cases}$$

To formalize these subproblems we introduce the following notation:

Notation 8.29. *Let* $\mathcal{A}^{fix}, \mathcal{A}^{miss} \subseteq \mathcal{A}_{change}$ *such that*

$$\mathcal{A}^{fix} \cap \mathcal{A}^{miss} = \emptyset.$$

Define $P(\mathcal{A}^{fix}, \mathcal{A}^{miss})$ *as an instance of (TDM-const) in which all connections* $a \in \mathcal{A}^{fix}$ *need to be maintained, and all connections* $a \in \mathcal{A}^{miss}$ *need not be considered.*

A similar notation will be used for (TDM-A), (TDM-B), and (TDM-C) in Section 8.6. Here, $P(\mathcal{A}^{fix}, \mathcal{A}^{miss})$ is given as the following integer program.

$(P(\mathcal{A}^{fix}, \mathcal{A}^{miss}))$

$$\min f_{P(\mathcal{A}^{fix},\mathcal{A}^{miss})} = \sum_{i \in \mathcal{E}} w_i y_i + \sum_{\substack{a \in \mathcal{A}_{change} \setminus \\ (\mathcal{A}^{miss} \cup \mathcal{A}^{fix})}} w_a T \bar{z}_a$$

such that

$$\begin{aligned} y_i \geq d_i & \quad \text{for all } i \in \mathcal{E}_{del} \\ y_i - y_j \leq s_a & \quad \text{for all } a = (i,j) \in \mathcal{A}^{fix} \cup \mathcal{A}_{wait} \cup \mathcal{A}_{drive} \\ -M \bar{z}_a + y_i - y_j \leq s_a & \quad \text{for all } a = (i,j) \in \mathcal{A}_{change} \setminus \left(\mathcal{A}^{miss} \cup \mathcal{A}^{fix}\right) \quad (8.47) \\ y_i \in \mathbb{N} & \quad \forall i \in \mathcal{E} \\ \bar{z}_a & \in \{0,1\} \quad \text{for all } a \in \mathcal{A}_{change} \setminus \left(\mathcal{A}^{fix} \cup \mathcal{A}^{miss}\right)) \end{aligned}$$

We will also consider the strong formulation ($P(\mathcal{A}^{fix}, \mathcal{A}^{miss})$-strong), in which constraint (8.47) is replaced by

$$-M_a \bar{z}_a + y_i - y_j \leq s_a \quad \text{for all } a = (i,j) \in \mathcal{A}_{change} \setminus \left(\mathcal{A}^{miss} \cup \mathcal{A}^{fix}\right).$$

Recall that $\mathcal{A}(\mathcal{A}^{fix}) = \mathcal{A}_{wait} \cup \mathcal{A}_{drive} \cup \mathcal{A}^{fix}$ and, to further simplify notation, define

$$\mathcal{A}^{open} = \mathcal{A}_{change} \setminus \left(\mathcal{A}^{fix} \cup \mathcal{A}^{miss}\right). \quad (8.48)$$

The following observations are the basis for the branch and bound approach.

- $P(\emptyset, \emptyset) =$(TDM-const).
- Fixing $\bar{z}_a = 0$ for some $a \in \mathcal{A}^{open}$ in $P(\mathcal{A}^{fix}, \mathcal{A}^{miss})$ leads to the new problem $P(\mathcal{A}^{fix} \cup \{a\}, \mathcal{A}^{miss})$.
- Fixing $\bar{z}_a = 1$ for some $a \in \mathcal{A}^{open}$ in $P(\mathcal{A}^{fix}, \mathcal{A}^{miss})$ leads to the new problem $P(\mathcal{A}^{fix}, \mathcal{A}^{miss} \cup \{a\})$.
- $P(\mathcal{A}^{fix}, \mathcal{A}_{change} \setminus \mathcal{A}^{fix}) =$(TT($\mathcal{A}^{fix}$)), and hence can be solved efficiently by Algorithm 12.

This means by fixing variables we always obtain subproblems of the same type. For the objective function values we know the following.

Lemma 8.30. *Let $\mathcal{A}_1^{fix} \subseteq \mathcal{A}_2^{fix} \subseteq \mathcal{A}_{change}$, $\mathcal{A}_1^{miss} \subseteq \mathcal{A}_2^{miss} \subseteq \mathcal{A}_{change}$, and let f_1^* be the optimal objective value of $P(\mathcal{A}_1^{fix}, \mathcal{A}_1^{miss})$, and f_2^* be the optimal objective value of $P(\mathcal{A}_2^{fix}, \mathcal{A}_2^{miss})$. Then*

$$f_1^* \leq f_2^* + \sum_{a \in \mathcal{A}_2^{miss} \setminus \mathcal{A}_1^{miss}} w_a T.$$

Especially, if f^* is the optimal objective value of (TDM-const) then

$$f^* \leq f_1^* + \sum_{a \in \mathcal{A}_1^{miss}} w_a T.$$

Proof. Any feasible solution $(y^2, \bar{z}^2)$ of $P(\mathcal{A}_2^{fix}, \mathcal{A}_2^{miss})$ can be extended to a feasible solution of $P(\mathcal{A}_1^{fix}, \mathcal{A}_1^{miss})$ by defining $y^1 = y^2$ and

$$\bar{z}_a^1 = \begin{cases} 0 & \text{if } a \in \mathcal{A}_2^{fix} \setminus \mathcal{A}_1^{fix} \\ 1 & \text{if } a \in \mathcal{A}_2^{miss} \setminus \mathcal{A}_1^{miss} \\ \bar{z}_a^2 & \text{if } \mathcal{A}_{change} \setminus \left(\mathcal{A}_2^{fix} \cup \mathcal{A}_2^{miss}\right). \end{cases}$$

For the objective function values we obtain

$$\begin{aligned} f_{P(\mathcal{A}_1^{fix}, \mathcal{A}_1^{miss})}(y^1, \bar{z}^1) &= \sum_{i \in \mathcal{E}} w_i y_i^1 + \sum_{\substack{a \in \mathcal{A}_{change} \setminus \\ (\mathcal{A}_1^{miss} \cup \mathcal{A}_1^{fix})}} w_a T \bar{z}_a^1 \\ &= \sum_{i \in \mathcal{E}} w_i y_i^1 + \sum_{\substack{a \in \mathcal{A}_{change} \setminus \\ (\mathcal{A}_2^{miss} \cup \mathcal{A}_2^{fix})}} w_a T \bar{z}_a^1 + \sum_{a \in \mathcal{A}_2^{miss} \setminus \mathcal{A}_1^{miss}} w_a T \bar{z}_a^1 \\ &= f_{P(\mathcal{A}_2^{fix}, \mathcal{A}_2^{miss})}(y^2, \bar{z}^2) + \sum_{a \in \mathcal{A}_2^{miss} \setminus \mathcal{A}_1^{miss}} w_a T. \end{aligned}$$

Now let $y_*^2, \bar{z}_*^2$ be an optimal solution of $P(\mathcal{A}_2^{fix}, \mathcal{A}_2^{miss})$ and $y_*^1, \bar{z}_*^1$ its extension as above to a solution of $P(\mathcal{A}_1^{fix}, \mathcal{A}_1^{miss})$. Consequently,

$$\begin{aligned} f_1^* &\leq f_{P(\mathcal{A}_1^{fix}, \mathcal{A}_1^{miss})}(y_*^1, \bar{z}_*^1) \\ &= f_{P(\mathcal{A}_2^{fix}, \mathcal{A}_2^{miss})}(y_*^2, \bar{z}_*^2) + \sum_{a \in \mathcal{A}_2^{miss} \setminus \mathcal{A}_1^{miss}} w_a T \\ &= f_2^* + \sum_{a \in \mathcal{A}_2^{miss} \setminus \mathcal{A}_1^{miss}} w_a T. \end{aligned}$$

□

One upper bound can be calculated easily, namely, if all vehicles are forced to wait; i.e., all connections should be maintained. Then we set $\mathcal{A}^{fix} = \mathcal{A}_{change}$ and obtain that

$$f^* \leq \sum_{i \in \mathcal{E}} y_i(\mathcal{A}_{change}) + 0.$$

The idea now is to start with $\mathcal{A}^{fix} = \mathcal{A}^{miss} = \emptyset$. In each node of the branch and bound tree, one of the $\bar{z}_a$-variables is fixed either to one or to zero. In both cases, the set $\mathcal{A}^{open}$ of open variables $\bar{z}_a$ is decreased by one. If all variables of $\mathcal{A}_{change}$ are either fixed to one or to zero, the resulting problem is of type

(TT). Hence, the optimal solution y is integer and can be found efficiently, see Chapter 7.

The following reduction lemma is a generalization of Lemma 8.13.

Lemma 8.31. *Let $\mathcal{A}^{fix}, \mathcal{A}^{miss} \subseteq \mathcal{A}_{change}$ and $\mathcal{A}^{fix} \cap \mathcal{A}^{miss} = \emptyset$. Let $y = y(\mathcal{A}_{change} \setminus \mathcal{A}^{miss})$ be an optimal solution of $TT(\mathcal{A}_{change} \setminus \mathcal{A}^{miss})$. Then there exists an optimal solution $y^*, \bar{z}^*$ of $P(\mathcal{A}^{fix}, \mathcal{A}^{miss})$ such that*

- *For all $i \in \mathcal{E}$: If $y_i = 0$ then $y_i^* = 0$.*
- *For all $a = (i,j) \in \mathcal{A}_{change} \setminus \mathcal{A}^{miss}$: If $y_i = 0$ then $\bar{z}_a^* = 0$.*
- *$M_a = y_i - s_a$ for all $a \in \mathcal{A}_{change} \setminus (\mathcal{A}^{miss} \cup \mathcal{A}^{fix})$ is large enough for the strong formulation of $P(\mathcal{A}^{fix}, \mathcal{A}^{miss})$.*

Proof. This can be shown analogously to Lemma 8.13, taking into account Lemma 8.27. □

Note that decreasing M yields better lower bounds according to Theorem 8.26. The algorithm can now be given.

Algorithm 14: Branch and bound for (TDM-const)

Input: $\mathcal{N}, \mathcal{P}, w_p, d_i, s_a, T$, and accuracy ϵ.
Output: Feasible solution $(y, \bar{z})$ for (TDM-const) with objective value f, such that $|f - f^*| \leq \epsilon$, if f^* is the optimal objective value.
Step 0.

$$\mathcal{A}^{miss} = \emptyset,$$
$$\mathcal{A}^{fix^*} = \emptyset, \text{ best solution obtained so far}$$
$$f^{\mathrm{u}} = \sum_{i \in \mathcal{E}} y_i(\mathcal{A}_{change}) \text{ upper bound,}$$
$$f^{\mathrm{l}} = \text{ optimal value of (LP-TDM-const) lower bound}$$
$$\text{List} = \{P(\mathcal{A}^{fix^*}, \mathcal{A}^{miss})\} \text{ with lower bound } f^l_{P(\mathcal{A}^{fix^*}, \mathcal{A}^{miss})} = f^{\mathrm{l}}.$$

Step 1.
1. If List$= \emptyset$, stop:
   Exact optimal solution is $y^* = y(\mathcal{A}^{fix^*}), \bar{z}^* = \bar{z}(y^*)$.
2. $f^{\mathrm{l}} = \min\{f_P^l : P \in \text{List}\}$
3. If $f^{\mathrm{u}} - f^{\mathrm{l}} \leq \epsilon$ stop:
   ϵ optimal solution is $y^* = y(\mathcal{A}^{fix^u}), \bar{z}^* = \bar{z}(y^*)$.

Step 2. Choose $P = P(\mathcal{A}^{fix}, \mathcal{A}^{miss}) \in$ List with current lower bound f_P^l. Let $\mathcal{A}^{open} = \mathcal{A}_{change} \setminus (\mathcal{A}^{fix} \cup \mathcal{A}^{miss})$.
Step 3. **Reduction of P:**
1. Calculate $y = y(\mathcal{A}_{change} \setminus \mathcal{A}^{miss})$ by Algorithm 12.
2. For all $a = (i,j) \in \mathcal{A}^{open}$: If $y_i = 0$ then

$$\mathcal{A}^{fix} = \mathcal{A}^{fix} \cup \{a\},$$
$$\mathcal{A}^{open} = \mathcal{A}^{open} \setminus \{a\}.$$

3. Set $M_a = y_i - s_a$ for all $a = (i,j) \in \mathcal{A}^{open}$

Step 4. **Bounds**

1. Solve LP-$P(\mathcal{A}^{fix}, \mathcal{A}^{miss})$ or its dual and obtain an optimal solution $(y^l, \bar{z}^l)$ with objective value f_P^l.
2. Let

$$\mathcal{A}^{fix^u} = \mathcal{A}^{fix} \cup \{a \in \mathcal{A}^{open} : \bar{z}_a = 0\}$$
$$\mathcal{A}^{miss^u} = \mathcal{A}^{miss} \cup \{a \in \mathcal{A}^{open} : \bar{z}_a > 0\}$$

and let $y^u = y(\mathcal{A}^{fix^u}), \bar{z}^u = \bar{z}(y^u)$ be an optimal solution of $P(\mathcal{A}^{fix^u}, \mathcal{A}^{miss^u})$ with objective value f_P^u.

Step 5. **Pruning**

1. If $\bar{z}^l$ is integer, prune by optimality, i.e.,

$$\mathcal{A}^{fix^*} = \{a \in \mathcal{A}^{open} : \bar{z}_a = 0\} \cup \mathcal{A}^{fix} \text{ if } f_P^l < f^u\}$$
$$f^{\mathbf{u}} = \min\{f_P^l, f^{\mathbf{u}}\}$$
$$\text{List} = \text{List} \setminus \{P\}.$$

Goto 1.

2. If $f_P^l \geq f^{\mathbf{u}}$ prune by bound, i.e. $\text{List} = \text{List} \setminus \{P\}$.
Goto 1.
3. If $f_P^u < f^u$ set

$$f^u = f_P^u,$$
$$\mathcal{A}^{fix^*} = \mathcal{A}^{fix^u}.$$

Step 6. Choose a minimal $a \in \mathcal{A}^{open}$ such that $\bar{z}_a^l \notin \{0,1\}$.

$$P_a^0 = P\{\mathcal{A}^{fix} \cup \{a\}, \mathcal{A}^{miss}) \text{ with } f_{P_a^0}^l = f_P^l$$
$$P_a^1 = P\{\mathcal{A}^{fix}, \mathcal{A}^{miss} \cup \{a\}) \text{ with } f_{P_a^1}^l = f_P^l$$
$$\text{List} = \text{List} \cup \{P_a^0, P_a^1\}$$

Goto 1.

Now we turn our attention back to (TDM-const) if additionally the never-meet property holds. In this case, the problem can be solved efficiently in $O(|\mathcal{A}|)$ time. The reason for this consists of the two facts listed below. Recall Notation 8.19 (page 139) where we introduced $\mathcal{H}(i, \mathcal{A}^{fix})$ as the set of events that can be reached from $i \in \mathcal{E}$ by using only changing arcs contained in $\mathcal{A}^{fix}$.

- First, if we fix $\bar{z}_a = 1$ for some $a = (i,j)$, this means that we can set $y_{i'} = 0$ for all $i' \in \mathcal{H}(j, \mathcal{A}_{change})$ and that all subsequent connections can be maintained (Lemma 8.20).
- Secondly, the problem can be decomposed into at most $|\mathcal{A}_{change}|$ independent subproblems due to the following lemma.

Lemma 8.32. *Let $i, j \in \mathcal{E}$, $i \neq j$, and let $(y, \bar{z})$ be a feasible solution of (TDM-const) with $y_i > 0, y_j > 0$. If the never-meet property holds, exactly one of the following three cases occurs.*

- $\mathcal{H}(i, \mathcal{A}_{change}) \subseteq \mathcal{H}(j, \mathcal{A}_{change})$
- $\mathcal{H}(j, \mathcal{A}_{change}) \subseteq \mathcal{H}(i, \mathcal{A}_{change})$
- $\mathcal{H}(i, \mathcal{A}_{change}) \cap \mathcal{H}(j, \mathcal{A}_{change}) = \emptyset$.

Proof. Let $i \neq j$. Assume that

- neither $\mathcal{H}(i, \mathcal{A}_{change}) \subseteq \mathcal{H}(j, \mathcal{A}_{change})$
- nor $\mathcal{H}(j, \mathcal{A}_{change}) \subseteq \mathcal{H}(i, \mathcal{A}_{change})$,

i.e., $i \notin \mathcal{H}(j, \mathcal{A}_{change})$ and $j \notin \mathcal{H}(i, \mathcal{A}_{change})$. Then we want to show that $\mathcal{H}(i, \mathcal{A}_{change}) \cap \mathcal{H}(j, \mathcal{A}_{change}) = \emptyset$. Assume the contrary, i.e. take some minimal $k \in \mathcal{H}(i, \mathcal{A}_{change}) \cap \mathcal{H}(j, \mathcal{A}_{change})$. This means there exist two paths $p^i = (i, \ldots, k^i, k)$ from i to k and $p^j = (j, \ldots, k^j, k)$ from j to k; for an illustration see Figure 8.9.

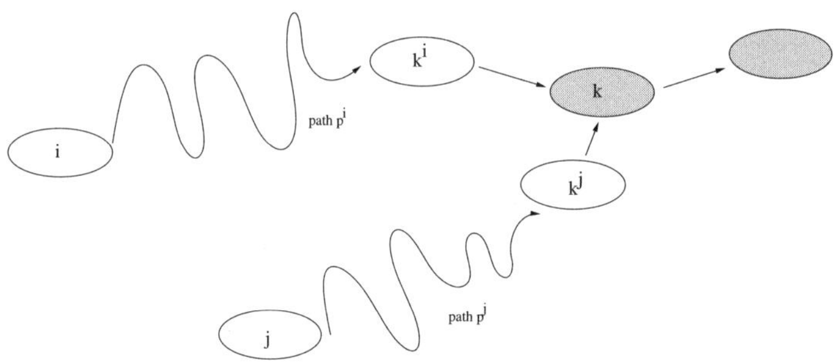

Fig. 8.9. The two paths p^i and p^j in the proof of Lemma 8.32.

Due to the minimality of k we conclude that

$$k^i \neq k^j.$$

Now let $y^0 = y^0(\mathcal{A}_{change})$. Then we know from Lemma 8.17 that $y^0_{j'} > 0$ for all $j' \in p_1 \cup p_2$, especially, $y^0_{k^i} > 0$ and $y^0_{k^j} > 0$, which is a contradiction to the never-meet property. □

For finding an (exact) optimal solution of (TDM-const) in case of the never-meet property we propose Algorithm 15. This algorithm will be given in the notation of event-activity networks, although this time it could have been stated as easily in the original notation. Suppose that some vehicle g has a delay at its arrival at station k. Then, independently of what we decide for later connections from this vehicle g to other vehicles, we can be sure that the vehicle will transfer its delay to subsequent stations, until it has been compensated by the slack times. This delay that will always be contributing to the objective will be formalized in the following more general notation.

Notation 8.33. *Let y be a time-minimal solution of (TDM) and $i \in \mathcal{E}$ with $y_i > 0$. Then denote*

$$F(i, y_i, \mathcal{A}^{fix}) = \sum_{j \in \mathcal{H}(i, \mathcal{A}^{fix})} w_j y_j.$$

The delay that will be transfered for sure by the delayed event i is then obtained by setting $\mathcal{A}^{fix} = \emptyset$. In the case that the event-activity network belongs to some PTN according to Notation 6.4 we get the following interpretation of $\mathcal{H}(i, \emptyset)$. Namely, it consists of a single path, describing all events belonging to the vehicle of event i. I.e.,

$$\mathcal{H}(i, \emptyset) = (i_1, i_2, \ldots, i_L),$$

and if y_i is given, the time-minimal solution y_{i_l} for all $l = 1, \ldots, L$ can be determined by

$$\begin{aligned} y_{i_1} &= y_i - s_{i_1, i} \\ y_{i_l} &= y_{i_{l-1}} - s_{i_{l-1}, i_l}. \end{aligned} \tag{8.49}$$

Before we formally state the algorithm, consider the following example, depicted in Figure 8.10.

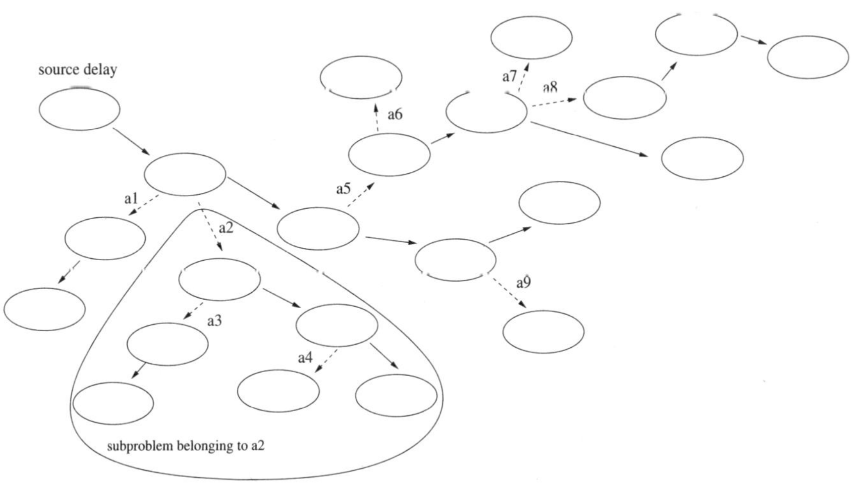

Fig. 8.10. Decomposition of $\mathcal{N}$ in case of the never-meet property. The changing activities are dashed.

The algorithm will first decompose P into subproblems which are collected in List(0). Each subproblem P_a is identified by a changing activity a. We obtain

$$\text{List}(0) = \{a_1, a_2, a_5, a_9\}.$$

The subproblem belonging to a_2 is depicted in Figure 8.10. To further decompose a subproblem P_a we store the subproblems belonging to its decomposition in List(a). In the example this gives the following lists:

$$\begin{aligned}\text{List}(a_2) &= \{a_3, a_4\}\\ \text{List}(a_5) &= \{a_6, a_7, a_8\},\end{aligned}$$

and the lists for all other $a \in \mathcal{A}_{change}$ are empty. All subproblems that might further be decomposed are stored in Decompose, and if a subproblem cannot further be decomposed it is collected in Compose. Hence, at the end of the decomposition step, we have

$$\text{Compose} = \{a_1, a_3, a_4, a_6, a_7, a_8, a_9\}.$$

Moreover, for each subproblem identified by changing activity a,

- maintain(a) contains the value of the objective function of the subproblem if a is maintained, and
- miss(a) contains the objective value if a is missed.

Algorithm 15: Enumeration for (TDM-const)

Input: $\mathcal{N}, \mathcal{P}, w_p, d_i, s_a, T$.
Output: Optimal solution of (TDM), if the never-meet property holds.
Step 0.
  1. Calculate w_a and w_i according to (8.25) and (8.24)
  2. Decompose $= \emptyset$, Compose $= \emptyset$.
  3. List(0) $= \emptyset$, $f(0) = 0$, $\bar{z}_a = 0$ for all $a \in \mathcal{A}_{change}$.
  4. For all $i \in \mathcal{E}_{del}$:
    a) Calculate $\mathcal{H}(i, \emptyset)$
    b) Calculate y_j for all $j \in \mathcal{H}(i, \emptyset)$ (by (8.49)).
    c) $f(0) = f(0) + F(i, d_i, \emptyset)$
    d) For all $a = (j_1, j_2) \in \mathcal{A}_{change}$ with $j_1 \in \mathcal{H}(i, \emptyset)$: If $y_{j_1} > 0$
      i. List(0) $=$ List(0) $\cup \{a\}$
      ii. Decompose $=$ Decompose $\cup \{a\}$
  5. If List(0) $= \emptyset$ stop: f is the optimal objective value, $\bar{z}_a = 0$ for all $a \in \mathcal{A}_{change}$.
Step 1. While Decompose $\neq \emptyset$
  1. Choose $a \in$ Decompose, $\tilde{a} \in \mathcal{A}_{change}$ with $a = (i_1, i_2) \in$ List($\tilde{a}$)
  2.

$$\begin{aligned}\text{List}(a) &= \emptyset,\\ \text{maintain}(a) &= 0,\\ \text{miss}(a) &= w_a T.\end{aligned}$$

3. $y_{i_2} = \max\{y_{i_1} - s_a, 0\}$
4. Calculate $\mathcal{H}(i_2, \emptyset)$
5. Calculate y_j for all $j \in \mathcal{H}(i_2, \emptyset)$ (by (8.49))
6. $\text{maintain}(a) = \text{maintain}(a) + F(i_2, y_{i_2}, \emptyset)$
7. For all $a' = (j_1, j_2) \in \mathcal{A}_{change}$ with $j_1 \in \mathcal{H}(i_2, \emptyset)$: If $y_{j_1} > 0$
   a) $\text{List}(a) = \text{List}(a) \cup \{a'\}$
   b) $\text{Decompose} = \text{Decompose} \cup \{a'\}$
8. If $\text{List}(a) = \emptyset$ then $\text{Compose} = \text{Compose} \cup \{a\}$.
9. $\text{Decompose} = \text{Decompose} \setminus \{a\}$.

Step 2. While $\text{Compose} \neq \emptyset$.
1. Choose $a \in \text{Compose}$, $\tilde{a} \in \mathcal{A}_{change}$ with $a \in \text{List}(\tilde{a})$
2. Define

$$\bar{z}_a = \begin{cases} 0 & \texttt{if } \text{maintain}(a) \leq \text{miss}(a) \\ 1 & \texttt{if } \text{maintain}(a) > \text{miss}(a) \end{cases}$$

$$f(a) = \min\{\text{maintain}(a), \text{miss}(a)\}$$

3.

$$\begin{aligned} \text{List}(\tilde{a}) &= \text{List}(\tilde{a}) \setminus \{a\} \\ \text{maintain}(\tilde{a}) &= \text{maintain}(\tilde{a}) + f(a) \\ \text{Compose} &= \text{Compose} \setminus \{a\} \end{aligned}$$

4. If $\text{List}(\tilde{a}) = \emptyset$ and $\tilde{a} \neq 0$ then $\text{Compose} = \text{Compose} \cup \{\tilde{a}\}$

Step 3.
1. For all $a \in \mathcal{A}_{change}$: If $\bar{z}_a = 1$ then set $\bar{z}_{a'} = 0$ for all $a' \neq a$ with $a \prec a'$.
2. Output: $f(\text{maintain}(0)), \bar{z}$.

Theorem 8.34. *Algorithm 15 is correct and runs in time $O(|\mathcal{A}|)$.*

Proof. We show by induction that at the end of Algorithm 15 $f(a)$ contains the objective value for the subproblem P_a. P_a is defined as (TDM-const) in the following network determined by $a = (i, j)$ and some delay y_i:

$$\mathcal{N}_a = (\mathcal{H}(i, \mathcal{A}_{change}), \mathcal{A}(\mathcal{H}(i, \mathcal{A}_{change})),$$

where

$$\mathcal{A}(\mathcal{H}(i, \mathcal{A}_{change})) = \{(j_1, j_2) \in \mathcal{A} : j_1, j_2 \in \mathcal{H}(i, \mathcal{A}_{change})).$$

The network belonging to P_{a_2} in the example is depicted in Figure 8.10.

Start: Let $a = (i, j)$ be a maximal element of $\mathcal{A}_{change}$ (with respect to $\prec$) and let $\mathcal{H} = \mathcal{H}(i, \mathcal{A}_{change})$. The subproblem with respect to a is (TDM-const) in the small network $\mathcal{N}_a = (\mathcal{H}, \mathcal{A}(\mathcal{H}))$. Since a is maximal, $\mathcal{A}(\mathcal{H})$ does not contain any changing activity. This means, $\text{List}(a) = \emptyset$ in step 2 of the algorithm. Furthermore,

$$\text{maintain}(a) = \sum_{i' \in \mathcal{H}} y_{i'} w_{i'}, \text{ and}$$
$$\text{miss}(a) = T w_a$$

give the objective values of this small network when maintaining or not maintaining activity a. To see the correctness of miss(a) we note that due to Lemma 8.20 $y_{i'} = 0$ for all $i' \in \mathcal{H}$. Since $a \in$ Compose we compare both values maintain(a) and miss(a) in step 3, and choose the better as (correct) objective value, which is then stored in $f(a)$.

Conclusion: Now take any $a = (i, j)$ and let the induction hypothesis be true for all a' with $a \prec a'$. Let $\mathcal{H} = \mathcal{H}(i, \emptyset)$ in this case. Then, if a is not maintained, we know from Lemma 8.20 that all connections $a' \in \mathcal{H}$ are maintained and all $i' \in \mathcal{H}$ satisfy $y_{i'} = 0$, i.e., the objective value is in this case given by miss(a) as calculated in step 2. For maintaining activity a the algorithm calculates in step 2 the delay which will be gained for sure, i.e., the delay of all events $i' \in \mathcal{H}$ that can be reached without passing any other changing activity, and stores it in maintain(a). All changing activities a' that can be reached from j without passing any other changing activity are stored in List(a). Each of these activities a' forms an independent subproblem on the smaller network $\mathcal{N}_{a'}$, since for $a_1 = (i_1, j_1), a_2 = (i_2, j_2) \in \text{List}(a)$ we have that

$$\mathcal{H}(i_1, \mathcal{A}_{change}) \cap \mathcal{H}(i_2, \mathcal{A}_{change}) = \emptyset$$

according to Lemma 8.32.

In step 3, we add up

$$\text{maintain}(a) + \sum_{a' \in \text{List}(a)} f(a').$$

This sum contains the best possible objective value for (TDM-const) on $\mathcal{N}_a$, if a is maintained, since $f(a')$ contains the optimal objective value of subproblem (TDM-const) on $\mathcal{N}_{a'}$ due to the induction hypothesis. Again, comparing the above sum (stored in maintain(a)) with miss(a) and choosing the smaller of both gives the best possible choice for activity a assuming the delay y_i as given.

Finally, in step 0, the problem with the given source delays is decomposed into a set of subproblems, given in List(0). All these subproblems are independent due to Lemma 8.32, and they are all solved optimally due to the Claim above.

Adding up these optimal values and adding the delay of all events which are reached before entering one of the subproblems gives the optimal objective function value $f(0)$.

For the time complexity we see that the number of subproblems equals the number of changing activities. For the decomposition step we have to process each activity exactly once, and in the composition step we need one comparison and one summation for each subproblem. The overall time complexity is hence linear in $|\mathcal{A}|$. □

The algorithm relies on the fact that each activity $a \in \mathcal{A}_{change}$ appears in exactly one list, i.e., for each $a \in \mathcal{A}_{change}$ there exists a unique $\tilde{a}$ such that $a \in \text{List}(\tilde{a})$, or $a \in \text{List}(0)$. If the never-meet property is not satisfied, this need not be the case, and hence Algorithm 15 cannot be applied to (TDM) for general problems. To resolve this problem (and to obtain a heuristic by applying Algorithm 15) one can either allow that the same element is added more than once to *Compose* in step 2 (this would mean to duplicate activities until the never-meet property is satisfied), or to update the values of *maintain* to the larger one, if an element which is already contained is added.

8.6 Solving the Total Delay Management Problem

Finally, we discuss how to solve the general version of (TDM). We first discuss lower and upper bounds and then put all obtained results together in a branch and bound procedure. For the following let f^* denote the optimal objective function value of (TDM) which is the same in all three formulations (TDM-A), (TDM-B), and (TDM-C).

Before we present an exact algorithm for solving (TDM), we suggest two heuristic approaches. Let us first focus on the formulation of (TDM-C). First note that any set $\mathcal{A}^{fix} \subseteq \mathcal{A}_{change}$ can be used to derive a feasible solution by the following procedure.

Algorithm 16: Calculating a feasible solution of (TDM)

Input: $\mathcal{N}, \mathcal{P}, w_p, d_i, s_a, T$, and $\mathcal{A}^{fix} \subseteq \mathcal{A}_{change}$.
Output: Reduced feasible solution for (TDM) with objective f.
Step 1. Determine $y(\mathcal{A}^{fix})$ by Algorithm 12.
Step 2. Calculate $C(y) = (\bar{z}, \tilde{z}, w)$ according to Notation 8.6.
Step 3. Output: $(y, \bar{z}, \tilde{z}, w)$ with objective value $f = f_{\text{TDM-C}}(y, \bar{z}, \tilde{z}, w)$

Notation 8.35. *Let $\mathcal{A}^{fix} \subseteq \mathcal{A}_{change}$. By*

$$R(\mathcal{A}^{fix}), f(\mathcal{A}^{fix})$$

denote the output of Algorithm 16 with input $\mathcal{A}^{fix}$, i.e.,

$$\begin{aligned} R(\mathcal{A}^{fix}) &= (y(\mathcal{A}^{fix}), C(y(\mathcal{A}^{fix}))) \\ f(\mathcal{A}^{fix}) &= f_{\text{TDM-C}}(R(\mathcal{A}^{fix})). \end{aligned}$$

All solutions of Algorithm 16 provide upper bounds on (TDM).

Lemma 8.36.

- $f^* \leq f(\mathcal{A}^{fix})$ *for all* $\mathcal{A}^{fix} \subseteq \mathcal{A}_{change}$.
- $f^* = \min_{\mathcal{A}^{fix} \subseteq \mathcal{A}_{change}} f(\mathcal{A}^{fix})$.

Proof. Part 1 follows immediately, since the solution $R(\mathcal{A}^{fix})$ is feasible for (TDM-C), see Lemma 8.7. For part 2, take an optimal solution $(y, \bar{z}, \tilde{z}, w)$ of (TDM-C) and define $\mathcal{A}^{fix} = \mathcal{A}^{fix}(\bar{z}) = \{a \in \mathcal{A} : \bar{z}_a = 0\}$. Then $R(\mathcal{A}^{fix}) = R^C(y, \bar{z}, \tilde{z}, w)$ (see Definition 8.9 on page 131), and hence

$$f(\mathcal{A}^{fix}) = f_{\text{TDM-C}}(R^C(y, \bar{z}, \tilde{z}, w)) \leq f_{\text{TDM-C}}(y, \bar{z}, \tilde{z}, w) = f^*$$

due to Theorem 8.11. □

Two obvious bounds can be calculated right from the start, namely, if all connections are maintained ($\mathcal{A}^{fix} = \mathcal{A}_{change}$), or if all vehicles depart on time ($\mathcal{A}^{fix} = \emptyset$). The question now is, how to choose a "good" set of fixed connections $\mathcal{A}^{fix}$ heuristically. Before we propose two heuristics, we introduce the following notation.

Notation 8.37. *Let $\mathcal{A}^{fix} \subseteq \mathcal{A}_{change}$. Denote*

$$W(i, \mathcal{A}^{fix}) = \sum_{p : p \cap \mathcal{H}(i, \mathcal{A}^{fix}) \neq \emptyset} w_p$$

as the number of all customers who will pass a station in $\mathcal{H}(i, \mathcal{A}^{fix})$.

If $i = (g, v, \text{dep})$ is the departure of some vehicle g at some station v, then $W(i, \emptyset)$ can be interpreted as follows. It contains the number of customers who plan to use vehicle g anywhere later than station v, i.e., those customers who

- either plan to pass station v with vehicle g, or
- plan to get on vehicle g at station v or at another station later than v.

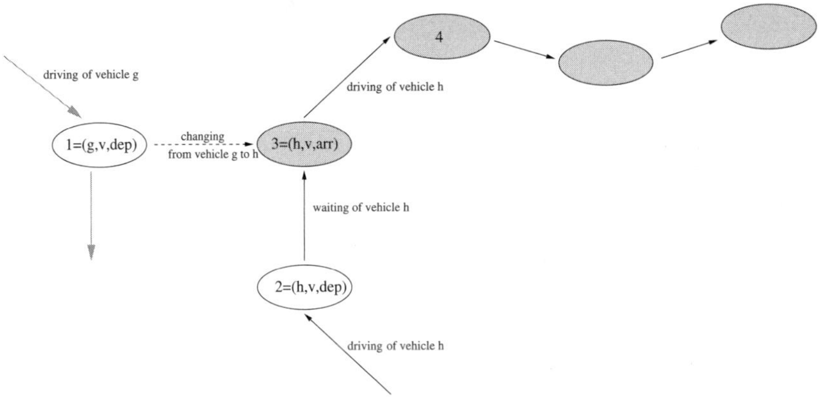

Fig. 8.11. Example with only one changing activity (dashed). The marked events belong to $\mathcal{H}(3, \emptyset)$

These are the customers who will most probably be annoyed if the vehicle waits at station v for some connecting vehicle.

The first heuristic is motivated by the following observation: Suppose the simple situation, depicted in Figure 8.11. We only have two vehicles g, h, and we assume that g has a delay of D minutes. We consider only one connection $\mathcal{U} = \{(g, h, v)\}$ and assume zero slack times. The network is represented as an event-activity network in Figure 8.11. To determine the optimal solution for this small problem, we first calculate

$$w_a = \sum_{p \in \mathcal{P}: a \in \mathcal{P}} w_p \text{ and}$$
$$W(3, \emptyset) = \sum_{p \in \mathcal{P}: p \cap \mathcal{H}(3,\emptyset) \neq \emptyset} w_p.$$

Note that $\mathcal{H}(3, \emptyset)$ exactly contains all stations which will be passed by vehicle h after station 2, and $W(3, \emptyset)$ sums up all customers who plan to use vehicle h at station 3 or later. (Note that all customers who get off vehicle h at station k will do this at event 2, and all customers who get on vehicle h at station k will do this at event 3 and will at least stay there until event 4.) Then,

- $\bar{z}_a = 1$ leads to an objective function value of $w_a T$
- $\bar{z}_a = 0$ leads to an objective function value of $W(3, \emptyset) D$.

Consequently, the optimal solution is given by

$$\bar{z}_a = 0 \Longleftrightarrow D \leq T \frac{w_a}{W(3, \emptyset)}.$$

This result is used in the first heuristic.

Heuristic 17: A-priori Heuristic for finding $\mathcal{A}^{fix}$

```
Input: 𝒩, 𝒫, w_p, d_i, s_a, T, and parameter Ã ⊆ 𝒜_change.
Output: 𝒜^fix.
Step 1. Calculate y = y(𝒜_change) by Algorithm 12.
Step 2. For all a = (i,j) ∈ 𝒜_change
     1. Calculate w_a = Σ_{p∈𝒫:a∈p} w_p
     2. Calculate W(j,Ã) = Σ_{p:p∩ℋ(j,Ã)≠0} w_p
     3. If y_i ≤ T w_a/W(j,∅) set z̄_a = 0,
        Otherwise set z̄_a = 1.
Step 3. Output: 𝒜^fix = {a ∈ 𝒜_change : z̄_a = 0}
```

Note that as in the example before, w_a contains the number of customers who wish to use changing activity a, and $W(j, \tilde{A})$ contains the number of customers who may be affected by a delay of event j. For $\tilde{A} = \emptyset$ we underestimate the number of these passengers (at least for small slack times), since we do not take into account that the delay might spread out along the changing activities. Increasing $\tilde{A}$ is a rather pessimistic approximation of the effects of waiting for other vehicles and will lead to less maintained connections. Heuristic 17 can be improved, if the current delay and the number of customers involved are updated after each decision made.

The next heuristic is based on the idea to maintain such connections, which are used by many customers.

Heuristic 18: Large-weight-Heuristic for finding $\mathcal{A}^{fix}$

```
Input: 𝒩, 𝒫, w_p, d_i, s_a, T, and percentage 0 ≤ p ≤ 1.
Output: 𝒜^fix.
Step 1. Sort the set of paths 𝒫 = {p_1, p_2, ..., p_|𝒫|}
    such that w_{p_1} ≥ w_{p_2} ≥ ... ≥ w_{p_|𝒫|}.
    Let W = Σ_{p∈𝒫} w_p.
Step 2. Find a number K such that
```

$$\sum_{k=1,\dots,K-1} w_p \leq pW \text{ and } \sum_{k=1,\dots,K} \geq pW.$$

```
Step 3. Output: 𝒜^fix = {a ∈ 𝒜_change : a ∈ p_k for some 1 ≤ k ≤ K}.
```

Note that an upper bound on (TDM) can also be obtained by

- solving (TDM-const) according to Lemma 8.15
- solving (TDM-const-zero) according to Lemma 8.24.

Now we turn our attention to the calculation of lower bounds for (TDM). To this end, we use the LP-relaxation of (TDM-B) (see page 124) to obtain a lower bound by linear programming. To strengthen this bound, we can decrease M as much as possible without making the optimal solution infeasible. As in the strong formulation of the LP-relaxation of (TDM-const) (see page 151) we replace M by a value M_a for all changing activities. Instead of (TDM-B) we obtain

(TDM-B-strong)

$$\min f_{\text{TDM-B}} = \sum_{p \in \mathcal{P}} w_p(q_p + T z_p)$$

such that

$$\begin{aligned} y_i &\geq d_i \quad \text{for all } i \in \mathcal{E}_{del} \\ y_i - y_j &\leq s_a \quad \text{for all } a = (i,j) \in \mathcal{A}_{wait} \cup \mathcal{A}_{drive} \\ -M_a z_p + y_i - y_j &\leq s_a \quad \text{for all } a = (i,j) \in \mathcal{A}_{change} \\ -M_p z_p + y_{i(p)} - q_p &\leq 0 \quad \text{for all } p \in \mathcal{P} \\ q_p &\geq 0 \quad \text{for all } p \in \mathcal{P} \\ y_i &\in \mathbb{N} \quad \text{for all } i \in \mathcal{E} \\ z_p &\in \{0,1\} \quad \text{for all } p \in \mathcal{P}. \end{aligned} \tag{8.50}$$

The following lemma is similar to Lemma 8.27.

Lemma 8.38. *Let $y = y(\mathcal{A}_{change})$ and let $M \geq M_a \geq y_i - s_a$ for all $a \in \mathcal{A}_{change}$ and $M \geq M_p \geq y_{i(p)}$ for all $p \in \mathcal{P}$. Then (TDM-B-strong) and (TDM-B) are equivalent.*

Proof. It is clear that the feasible set of (TDM-B-strong) is contained in (TDM-B), since $M \geq D \geq M_a$ for all $a \in \mathcal{A}_{change}$. This means that each feasible solution of (TDM-B-strong) is also a feasible solution of (TDM-B) with the same objective value.

On the other hand, take an optimal solution (y^*, z^*, q^*) of (TDM-B). Due to Lemma 8.3 we can assume that y^* is time-minimal. Furthermore, note that $\{a \in \mathcal{A}_{change} : y_i^* - y_j^* \leq s_a\} \subseteq \mathcal{A}_{change}$, hence we know from part 2 of Lemma 7.8 that

$$y^* \leq y = y(\mathcal{A}_{change}).$$

This gives us for all $a \in \mathcal{A}_{change}$ that

$$\begin{aligned} M_a &\geq y_i - s_a \ \geq \ y_i^* - y_j^* - s_a, \text{ and} \\ M_p &\geq y_{i(p)} \ \geq \ y_{i(p)}^* - q_p^* \end{aligned}$$

hence (y^*, z^*, u^*) is feasible for (TDM-B-strong) with the same objective function value and the result follows. □

We conclude that in fact, the LP-relaxation of (TDM-B-strong) yields better bounds than (LP-TDM-B). This is formulated in the next result.

Lemma 8.39. *Let $M_a \leq M'_a$ for all $a \in \mathcal{A}_{change}$, $M_p \leq M'_p$ for all $p \in \mathcal{P}$, and*

- *let f be the optimal solution of (LP-TDM-B-strong) with respect to M_a, M_p and*
- *f' be the optimal solution of (LP-TDM-B-strong) with respect to M'_a, M'_p.*

Then $f \geq f'$.

Proof. The lemma follows from the fact that (LP-TDM-strong) with respect to M'_a, M'_p is a relaxation of (LP-TDM-strong) with respect to M_a, M_p. □

The idea of the branch and bound approach is similar to that of Algorithm 14. In each step we consider the earliest unknown connection $a = (i, j)$ with $y_i > 0$ and define two subproblems by either fixing the respective variable $\bar{z}_a = 0$, or allowing $\bar{z}_a = 1$. The subproblems we obtain are the same as in Notation 8.29 (on page 153). More precisely, we consider subproblems of type

$$P(\mathcal{A}^{fix}, \mathcal{A}^{miss})$$

defined as follows:

- If we consider (TDM-C) we add

$$y_i - y_j \leq s_a \text{ for all } a \in \mathcal{A}^{fix} \tag{8.51}$$

  to the formulation of (TDM-C), and omit the constraints

$$-M\bar{z}_a + y_i - y_j \leq s_a \text{ for all } a \in \mathcal{A}^{miss}. \tag{8.52}$$

- If we deal with (TDM-A) or (TDM-B) we again add constraints (8.51) to the respective formulation, but instead of (8.52) we determine the paths containing a connection of $\mathcal{A}^{miss}$, i.e.,

$$\mathcal{P}^{miss} = \{p \in \mathcal{P} : p \cap \mathcal{A}^{miss} \neq \emptyset\}$$

  and omit the constraints

$$-Mz_p + y_i - y_j \leq s_a \text{ for all } p \in \mathcal{P}^{miss}. \tag{8.53}$$

In the algorithm we will use the following two types of reduction procedures for the networks appearing in the subproblems.

Late reduction: The concept has been introduced in Lemmas 8.13 and 8.31 for the subproblems, see also Notation 8.14 on page 133. Late reduction means that we can delete events and activities that will never be delayed in any time-minimal solution.

Early reduction is relevant, if in a subproblem the first K connections have already been fixed either to be maintained, or to be not maintained. Then all events before the first open connection will not be influenced by the wait-depart decision of the remaining connections later on. This means that we can delete all events and all activities appearing before we reach a minimal open connection.

Moreover, the set of paths can be reduced since paths which have no events in common with the remaining network, or are already known to be missed, need not be considered any more, and the constants M_a, M_p needed for solving the relaxations can be adapted. The reduction works as follows.

Algorithm 19: Reduction of (TDM)($\mathcal{A}^{fix}, \mathcal{A}^{miss}$)

Input: $\mathcal{N}, \mathcal{P}, w_p, d_i, s_a, T$, current objective value f_P, and $\mathcal{A}^{fix}, \mathcal{A}^{miss}$.
Output: Reduced data of (TDM) problem, updated f_P.
Step 1: Calculate $y = y(\mathcal{A}_{change} \setminus \mathcal{A}^{miss})$.
Step 2: Early reduction
1. $\mathcal{A}^{open} = \mathcal{A}_{change} \setminus (\mathcal{A}^{fix} \cup \mathcal{A}^{miss})$.
2. Define the new set of delayed events $\mathcal{E}_{del}$ as the minimal elements of the set

$$\{i \in \mathcal{E} : y_i > 0 \text{ and there exists } a = (i, j) \in \mathcal{A}^{open}\},$$

each of them with source delay y_i.
3. Redefine $\mathcal{E}$. For all $i \in \mathcal{E}$: If for **no** $j \in \mathcal{E}_{del}$ $j \prec i$, then delete i from $\mathcal{E}$.
4. Redefine $\mathcal{A}$. For all $a = (i_1, i_2) \in \mathcal{A}$: If there does **not** exist $j \in \mathcal{E}_{del}$ with $j \prec i_1$ then
   a) delete a from $\mathcal{A}$.
   b) if $a \subset \mathcal{A}^{open}$:
      - $\mathcal{A}^{fix} = \mathcal{A}^{fix} \cup \{a\}$
      - $\mathcal{A}^{open} = \mathcal{A}^{open} \setminus \{a\}$

Step 3: Late reduction
1. For all $a \in \mathcal{A}^{open}$: If $y_i = 0$ then

$$\mathcal{A}^{fix} = \mathcal{A}^{fix} \cup \{a\},$$
$$\mathcal{A}^{open} = \mathcal{A}^{open} \setminus \{a\}.$$

2. Set $M_a = y_i - s_a$ for all $a \in \mathcal{A}_{change}$, $M_p = y_{i(p)}$ for all $p \in \mathcal{P}$.

Step 4: Redefine $\mathcal{P}$: For all $p \in \mathcal{P}$:
1. If $p \cap \mathcal{A}^{miss} \neq \emptyset$ let $f_P = f_P + w_p T$ and delete p from $\mathcal{P}$.
2. If $p \cap \mathcal{E} = \emptyset$ let $f_P = f_P + w_p y_{i(p)}$ and delete p from $\mathcal{P}$.

The more decisions are made, the more reductions can be made, and the chance that the problem can be decomposed into independent subproblems increases.

Notation 8.40. *Let $\tilde{a} = (\tilde{i}, \tilde{j})$ be a minimal element of $\mathcal{A}^{open}$, i.e., all $a \prec \tilde{a}$ are either in $\mathcal{A}^{fix}$ or in $\mathcal{A}^{miss}$. The wait-depart decision of $\tilde{a}$ is called* **independent**, *if*

$$\mathcal{H}(\tilde{i}, \mathcal{A}^{open} \cup \mathcal{A}^{fix}) \cap \mathcal{H}(i, \mathcal{A}^{open} \cup \mathcal{A}^{fix}) = \emptyset$$

for all i satisfying the following two conditions.

- $y_i(\mathcal{A}^{open} \cup \mathcal{A}^{fix}) > 0.$
- *Neither $\tilde{i} \prec i$ nor $i \prec \tilde{i}$.*

Note that all subproblems are independent if the never-meet property holds. In the next lemma we show that the optimal decision can often be determined directly, if it is independent. This result is utilized in step 8 of the branch and bound algorithm.

Lemma 8.41. *Let $\tilde{a} = (\tilde{i}, \tilde{j}) \in \mathcal{A}_{change}$ and let the wait-depart decision of $\tilde{a}$ be independent. Define*

$$\begin{aligned} f_{miss} &= T w_{\tilde{a}} \\ f^l_{maintain} &= \sum_{i \in \mathcal{H}(\tilde{i}, \emptyset)} w_i y_i(\mathcal{A}^{fix} \cup \{\tilde{a}\}) \\ f^u_{maintain} &= \sum_{i \in \mathcal{H}(\tilde{i}, \mathcal{A}^{open} \cup \mathcal{A}^{fix})} w_i y_i(\mathcal{A}^{open} \cup \mathcal{A}^{fix}) \end{aligned}$$

recalling the definition of w_a and w_i according to (8.25) and (8.24) on page 134 (in the network after performing the reduction algorithm).
For the optimal choice of $\bar{z}_{\tilde{a}}$ the following holds:

- *If $f_{miss} > f^u_{maintain}$ then $\bar{z}_{\tilde{a}} = 0$.*
- *If $f^l_{maintain} > f_{miss}$ then $\bar{z}_{\tilde{a}} = 1$.*

Proof. For this proof we use (TDM-A). We rewrite the objective to

$$\begin{aligned} f_{\text{TDM-A}} &= \sum_{p \in \mathcal{P}} w_p (y_{i(p)}(1 - z_p) + T z_p) \\ &= \sum_{p \in \mathcal{P}: i(p) \in \mathcal{H}(\tilde{i}, \mathcal{A}_{change})} w_p (y_{i(p)}(1 - z_p) + T z_p) + C \\ &= f(y, z) + \text{const}, \end{aligned}$$

where the term const is not affected by the decision concerning the connection $\tilde{a}$ due to the independence of the decision concerning $\tilde{a}$ (meaning that in no feasible solution does any other delayed vehicle reach $\mathcal{H}$).

- If (y, z) is an optimal solution of (TDM-A) with the additional constraint $y_{\tilde{i}} - y_{\tilde{j}} \leq s_{\tilde{a}}$, then denote $f_0 = f(y, z)$.

- Similarly, if (y, z) is the best possible solution satisfying $y_{\tilde{i}} - y_{\tilde{j}} > s_{\tilde{a}}$, then let $f_1 = f(y, z)$.

We show that

1. $f^l_{maintain} \leq f_0 \leq f^u_{maintain}$, and
2. $f_{miss} = f_1$,

hence yielding that for $f_{miss} > f^u_{maintain}$ we obtain $f_0 < f_1$ and hence $\bar{z}_a = 0$ is the optimal choice, and analogously for $f^l_{maintain} > f_{miss}$ $\bar{z}_a = 1$ occurs in the optimal solution.

$f_{miss} = f_1$: If connection $\tilde{a}$ is missed we obtain $z_p = 1$ for all $p \in \mathcal{P}$ with $\tilde{a} \in p$. Moreover, similarly as in Lemma 8.20 we know that all $y_i = 0$ for $\tilde{i} \prec i$ and hence all other paths ending in $\mathcal{H}(\tilde{i}, \mathcal{A}^{fix} \cup \mathcal{A}^{open})$ make no contribution to the objective function value, i.e.,

$$f_1 = \sum_{p \in \mathcal{P}: a \in p} w_p T = w_{\tilde{a}} = f_{miss}.$$

$f^l_{maintain} \leq f_0$: Now consider the case that $\tilde{a}$ is maintained. Then at least all customers getting out at a node $i \in \mathcal{H}(\tilde{i}, \emptyset)$ (of the same vehicle h belonging to $\tilde{j}$ later than $\tilde{a}$) gain a delay of at least y_i (since there is no changing activity except $\tilde{a}$ between $\tilde{i}$ and i. This gives us that

$$f_0 \geq \sum_{i \in \mathcal{H}(\tilde{i}, \emptyset)} w_i y_i(\{\tilde{a}\}) = f^l_{maintain}.$$

$f^u_{maintain} \geq f_0$: Finally, consider the solution of (TDM) in which $\tilde{a}$ and all subsequent connections are maintained. Then no customer (in the set of paths $\mathcal{P}$ after the early reduction) who planned to get off in $\mathcal{H}(\tilde{i}, \mathcal{A}^{open} \cup \mathcal{A}^{fix})$ misses his connection, but arrives with a delay of $y_{i(p)}$ for $y = y(\mathcal{A}^{open} \cup \mathcal{A}^{fix})$. We hence obtain

$$f_0 \leq \sum_{i \in \mathcal{H}(\tilde{i}, \mathcal{A}_{change})} w_i y_i(\mathcal{A}_{change}) = f^u_{maintain}.$$

□

Finally, we summarize the results of this section in the next algorithm. We start with the earliest decision a to be made and branch into $\bar{z}_a = 1$ and $\bar{z}_a = 0$. For both subproblems we calculate upper and lower bounds and also investigate their structure to apply Lemma 8.41 or solve them optimally in case of the never-meet property. The more decisions have been fixed the more likely an easy subproblem is obtained. To increase the efficiency of the algorithm we add both reductions steps.

Algorithm 20: Branch and bound for (TDM)

Input: $\mathcal{N}, \mathcal{P}, w_p, d_i, s_a, T$, and accuracy ϵ.
Output: Feasible solution $R(\mathcal{A}^{fix})$ of (TDM-B) with objective value $f = f(\mathcal{A}^{fix})$, such that $|f - f^*| \leq \epsilon$, if f^* is the optimal objective value.
Step 0.

$$\mathcal{A}^{miss} = \emptyset,$$
$$\mathcal{A}^{fix^*} = \emptyset, \text{ best solution obtained so far}$$
$$f^{\mathrm{u}} = \text{Upper bound calculated by Heuristic 17 or 18}$$
$$f^{\mathrm{l}} = \text{Lower bound calculated by LP-Relaxation of (TDM-B)}$$
$$\text{List} = \{P(\mathcal{A}^{fix^*}, \mathcal{A}^{miss})\} \text{ with current objective} f^l_{P(\mathcal{A}^{fix^*}, \mathcal{A}^{miss})} = 0$$
$$\text{and } f_{P(\mathcal{A}^{fix^*}, \mathcal{A}^{miss})} = 0$$

Step 1.
1. If List$= \emptyset$, stop.
   Calculate the exact optimal solution $R^{\mathcal{A}^{fix^*}}$ by Algorithm 16.
2. $f^{\mathrm{l}} = \min\{f^l_P : P \in \text{List}\}$
3. If $f^{\mathrm{u}} - f^{\mathrm{l}} \leq \epsilon$ stop:
   Calculate the ϵ optimal solution solution $R^{\mathcal{A}^{fix^*}}$ by Algorithm 16.

Step 2. Choose $P = P(\mathcal{A}^{fix}, \mathcal{A}^{miss}) \in \text{List}$ with current objective f_P.
Let $\mathcal{A}^{open} = \mathcal{A}_{change} \setminus \left(\mathcal{A}^{fix} \cup \mathcal{A}^{miss}\right)$.
Step 3. **Reduction of P** Perform Algorithm 19
Step 4: **Never-meet property** Check if the never-meet property is satisfied (Algorithm 13). If yes:
Solve P optimally by Algorithm 15 and let $f^l = f^u$ be the optimal solution and $\mathcal{A}^{fix^u}$ be the set of maintained connections.
Goto 7.
Step 5. **Lower bound** Solve the relaxation LP-TDM-B-strong($\mathcal{A}^{fix}, \mathcal{A}^{miss}$) and let f^l be its objective value and (y, z, u) its solution.
If (y, z, u) is integer, then
- $f^l = f^u$,
- $\mathcal{A}^{fix^u} = \{a \in \mathcal{A}^{open} : \bar{z}_a = 0\}$ and
- goto 6.

Step 6. **Upper bound** Choose at least one of the following steps.
1. Let $\mathcal{A}^{fix^u} = \{a \in \mathcal{A}_{change} : a \in p \text{ with } z_p = 0\}$ in the solution of LP-TDM-B($\mathcal{A}^{fix}, \mathcal{A}^{miss}$) in step 5 and calculate $f^u = f(\mathcal{A}^{fix})$ according to Algorithm 16.
2. Calculate w_a and w_i according to (8.25) and (8.24) Solve P with fixed weights by Algorithm 14. If the obtained results are better, update f^u, $\mathcal{A}^{fix^u}$.
3. Run Heuristic 17 or 18 for solving $P(\mathcal{A}^{fix}, \mathcal{A}^{miss})$. If the obtained results are better, update f^u, $\mathcal{A}^{fix^u}$.

Step 7. **Pruning**

1. If $f^l = f^u$, prune by optimality, i.e.,

$$\begin{aligned} \mathcal{A}^{fix*} &= \mathcal{A}^{fix^u} \cup \mathcal{A}^{fix} \texttt{ if } f^l + f_P < f^{\mathbf{u}} \\ f^{\mathbf{u}} &= \min\{f^l + f_P, f^{\mathbf{u}}\} \\ \text{List} &= \text{List} \setminus \{P\}. \end{aligned}$$

Goto 1.

2. If $f^l > f^u$ or if $f_P + f^l \geq f^{\mathbf{u}}$ prune by bound, i.e.,

$$\text{List} = \text{List} \setminus \{P\}.$$

Goto 1.

3. If $f^u + f_P < f^{\mathbf{u}}$ set

$$\begin{aligned} f^{\mathbf{u}} &= f^u + f_P \\ \mathcal{A}^{fix*} &= \mathcal{A}^{fix^u} \cup \mathcal{A}^{fix}. \end{aligned}$$

Step 8. Choose a minimal $a \in \mathcal{A}^{open}$.

$$\begin{aligned} P_a^0 &= P(\mathcal{A}^{fix} \cup \{a\}, \mathcal{A}^{miss}) \\ P_a^1 &= P(\mathcal{A}^{fix}, \mathcal{A}^{miss} \cup \{a\}) \texttt{ with} \\ f_{P_a{}^0} &= f_{P_a{}^1} = f_P \texttt{ and} \\ f^l_{P_a{}^0} &= f^l_{P_a{}^1} = f_P + f^l. \end{aligned}$$

1. If the wait depart-decision of a is independent:
   - If $f_{miss} \geq f^u_{maintain}$ set $\text{List} = \text{List} \cup \{P_a^0\}$
   - If $f^l_{maintain} \geq f_{miss}$ set $\text{List} = \text{List} \cup \{P_a^1\}$
2. Otherwise, $\text{List} = \text{List} \cup \{P_a^0, P_a^1\}$

Goto 1.

Since the choice of the next changing activity is performed according to $\prec$ step 3 can be performed very efficiently if the reduced data of the parent node has been stored. Note that step 8 of the algorithm is due to Lemma 8.41.

9

The Bicriteria Delay Management Problem

All models developed in the previous chapter assume that we know the exact paths the customers would like to use. If this is not the case, we at least need the OD-matrix W, such that we can estimate the paths of the customers. Moreover, the OD-matrix should also provide the time at which a customer starts his journey. While an OD-matrix sometimes is known for railway transportation, many bus companies do not even have this information, and time-dependent OD matrices are usually not available at all. In particular, nearly nothing is known about the customers who change between different transportation companies. In this chapter we hence suggest an approach to the delay management problem not requiring all this information. To this end we deal with the bicriteria delay management problem (BDM) of minimizing the number of missed connections, and at the same time, minimizing the sum of all delays over all vehicles.

(BDM)

Given PTN, $F, \mathcal{U}$, *minimal necessary times for driving, waiting, and changing, a feasible timetable* (Π_{arr}, Π_{dep}), *a set of weighted paths* $\mathcal{P}$ *through* PTN, *and a set of delayed events* $\mathcal{E}_{del}$, *find a perturbed feasible timetable* (x_{arr}, x_{dep}), *such that both*

$$f_{\mathcal{E}} = \sum_{g \in F, v \in V^g} x_{arr}{}_g^v - \Pi_{arr}{}_g^v, \text{ and}$$

$$f_{\mathcal{A}} = \sum_{\substack{(g,h,v) \in \mathcal{U}: \\ (g,h,v) \text{ is missed}}} w_{ghv}$$

are minimized.

Again, note that for this approach we do not need any detailed data about the customers, but only require a rough estimation to specify the importance of the connections for the customers. To model this problem we again use the concept of event-activity networks as introduced in Section 6.4.

Chapter 9 is structured as follows: We first show that (BDM) is NP-hard and then present an integer programming formulation of the problem. For this problem we are able to show that *all* Pareto solutions are time-minimal. We point out that finding supported Pareto solutions is equivalent to an instance of (TDM-const). To find all Pareto solutions of (BDM), we then develop an exact algorithm based on a procedure for the discrete time/cost tradeoff problem.

9.1 A First Analysis

Our first result clarifies the complexity status of (BDM).

Theorem 9.1. *(BDM) is NP-hard, even if all slack times are zero, and no two connections are contained in the same connected component of the given* PTN.

Proof. We reduce (BDM) to the knapsack problem, which is NP-hard, see [GJ79a]. Given an instance of the knapsack problem, i.e., n items, each of them with a cost $c_i \geq 0$ and a benefit $b_i \geq 0$, $i = 1, \ldots, n$, and threshold parameters C and B, does there exist a subset of items with total weight less than or equal to C and total benefit at least B? For our proof we may assume that $b_i > 0$, since items with benefit $b_i = 0$ will never be chosen in an optimal solution of the knapsack problem.

To construct an instance of (BDM) from the knapsack problem we define PTN $= (V, E)$ as follows: V consists of $3n$ nodes, numbered by v_{1i}, v_{2i}, v_{3i}, $i = 1, \ldots, n$ and

$$E = \{(v_{1i}, v_{2i}), (v_{2_i}, v_{3i}) : i = 1, \ldots, n\}.$$

Consider $2n$ vehicles (trains) $t_1, \ldots, t_n, t'_1, \ldots, t'_n$ where t_i goes from v_{1i} to v_{2i}, while t'_i starts at v_{2i} and arrives at v_{3i}, see Figure 9.1. We assume zero slack times, i.e., $s_g^v = s_{gh}^v = s_g^{vu} = 0$ for all $g, h \in F, k, u \in V$. We set the weights for the connections from t_i to t'_i at v_{2i} as b_i, i.e.,

$$w_{t_i t'_i v_{2i}} = b_i \text{ for all } i \in \{1, \ldots, n\}.$$

Furthermore, assume that the arrival of t_i at station v_{2i} is delayed by c_i for all $i = 1, \ldots, n$, and we set

$$C' = \sum_{i=1}^{n} c_i + C \text{ and}$$

$$B' = \sum_{i=1}^{n} b_i - B.$$

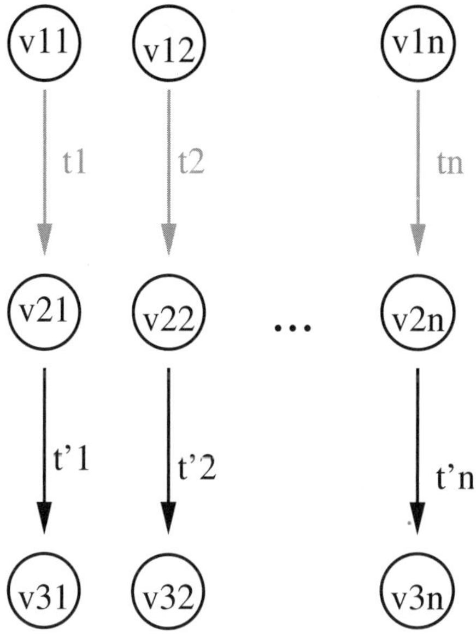

Fig. 9.1. Reduction of (BDM) to the knapsack problem.

Claim: There exists a solution to (BDM) with $f_{\mathcal{E}} \leq C'$ and $f_{\mathcal{A}} \leq B'$ if and only if the instance of the knapsack problem can be answered by yes.

To see this equivalence, let x_{arr}, x_{dep} be a solution of (BDM) and define

$$\mathcal{U}^* := \{i : x_{dep}{}_{t'_i}^{v_{2i}} \geq x_{arr}{}_{t_i}^{v_{2i}} + L_{t_i t'_i}^{v_{2i}}\}$$

as the set of maintained connections, each of them corresponding to an item packed into the rucksack. Then we obtain

$$\begin{aligned} & f_{\mathcal{E}}(x_{arr}, x_{dep}) \leq C' \\ \Longleftrightarrow & \sum_{i=1}^{n} x_{arr}{}_{t_i}^{v_{2i}} - \Pi_{arr}{}_{t_i}^{v_{2i}} + \sum_{i \in \mathcal{U}^*} x_{arr}{}_{t'_i}^{v_{3i}} - \Pi_{arr}{}_{t'_i}^{v_{3i}} \leq C' \\ \Longleftrightarrow & \sum_{i=1}^{n} c_i + \sum_{i \in \mathcal{U}^*} c_i \leq \sum_{i=1}^{n} c_i + C \\ \Longleftrightarrow & \sum_{i \in \mathcal{U}^*} c_i \leq C \end{aligned}$$

and

$$\begin{aligned}
& f_{\mathcal{A}}(x_{arr}, x_{dep}) \leq B' \\
\Longleftrightarrow & \sum_{i:(t_i,t'_i,v_{2i})\ missed} w_i \leq B' \\
\Longleftrightarrow & \sum_{i \notin \mathcal{U}^*} b_i \leq \sum_{i=1}^{n} b_i - B \\
\Longleftrightarrow & \sum_{i \in \mathcal{U}^*} b_i \geq B,
\end{aligned}$$

hence the claim is established. □

With the notation introduced in Section 6.4 we are in the position to restate the *bicriteria delay management problem* (BDM) as follows.

(BDM) Given $\mathcal{N} = (\mathcal{E}, \mathcal{A})$, a feasible timetable $\Pi_i, i \in \mathcal{E}$, minimal arc durations $L_a, a \in \mathcal{A}$, and source delays d_i determine a feasible perturbed timetable $x_i, i \in \mathcal{E}$ such that the following two objectives are minimized. The sum of all delays over all vehicles and all stations:

$$\sum_{i \in \mathcal{E}_{arr}} x_i - \Pi_i,$$

which is equivalent to the minimization of

$$f_{\mathcal{E}} = \sum_{i \in \mathcal{E}_{arr}} x_i$$

The weighted number of missed connections , given as

$$f_{\mathcal{A}} = \sum_{a=(i,j) \in \mathcal{A}_{change}: x_j - x_i < L_a} w_a.$$

We assume that $w_a > 0$ for all $a \in \mathcal{A}_{change}$, otherwise we simply delete a from $\mathcal{A}_{change}$.

What we mean by "minimizing simultaneously" is to find Pareto solutions of the problem with respect to $f_{\mathcal{E}}$ and $f_{\mathcal{A}}$. Recall from Appendix B that if x_1, x_2 denote two feasible perturbed timetables, then x_1 dominates x_2 if

$$\begin{aligned}
f_{\mathcal{E}}(x_1) &\leq f_{\mathcal{E}}(x_2) \ \text{ and} \\
f_{\mathcal{A}}(x_1) &\leq f_{\mathcal{A}}(x_2),
\end{aligned}$$

where at least one of the inequalities is strict. Then a *Pareto solution* is a feasible perturbed timetable which is not dominated by any other feasible perturbed timetable (see Appendix B). The points $\begin{pmatrix} f_{\mathcal{E}}(x^*) \\ f_{\mathcal{A}}(x^*) \end{pmatrix}$ in objective space are called *efficient* points, if x is a Pareto solution.

9.2 Integer Programming Formulation

We now formulate (BDM) as a bicriteria (linear) integer program. To this end, let $M > \max_{i \in \mathcal{E}} d_i$.

$$\min f_{\text{BDM}} = \begin{pmatrix} \sum_{i \in \mathcal{E}_{arr}} y_i \\ \sum_{a \in \mathcal{A}_{change}} w_a \bar{z}_a \end{pmatrix} \tag{9.1}$$

such that

$$y_i \geq d_i \quad \text{for all } i \in \mathcal{E} \tag{9.2}$$

$$y_i - y_j \leq s_a \quad \text{for all } a = (i,j) \in \mathcal{A}_{wait} \cup \mathcal{A}_{drive} \tag{9.3}$$

$$-M\bar{z}_a + y_i - y_j \leq s_a \quad \text{for all } a = (i,j) \in \mathcal{A}_{change} \tag{9.4}$$

$$y_i \in \mathbb{N} \quad \text{for all } i \in \mathcal{E}$$

$$\bar{z}_a \in \{0,1\} \quad \text{for all } a \in \mathcal{A}_{change}.$$

First we show that y is time-minimal in all Pareto solutions of (BDM). Recall that

$$\mathcal{A}^{fix}(\bar{z}) = \{a \in \mathcal{A}_{change} : \bar{z}_a = 0\}.$$

Lemma 9.2. *Let $(y, \bar{z})$ be a Pareto solution of (BDM). Then*

1. *$y = y(\mathcal{A}^{fix}(\bar{z}))$.*
2. *For all $a = (i,j) \in \mathcal{A}_{change}$:*

$$\bar{z}_a - 0 \Longleftrightarrow y_i - y_j \leq s_a.$$

Proof. 1. Denote $y^* = y(\mathcal{A}^{fix}(z))$ and assume that $y \neq y^*$. Since y is a feasible perturbed timetable we know from part 1 of Lemma 7.8 that $y^* \leq y$, and $y \neq y^*$ gives us additionally that there exists $i \in \mathcal{E}$ with $y_i^* < y_i$. Moreover, since y^* is feasible for $\text{TT}(\mathcal{A}^{fix}(\bar{z}))$ we conclude that $(y^*, \bar{z})$ also is a feasible solution of (BDM). Hence,

$$\begin{aligned} f_{\mathcal{E}}(y^*, \bar{z}) &= \sum_{i \in \mathcal{E}_{arr}} y_i^* \\ &< \sum_{i \in \mathcal{E}_{arr}} y_i = f_{\mathcal{E}}(y, z), \text{ and} \\ f_{\mathcal{A}}(y^*, \bar{z}) &= f_{\mathcal{A}}(y, \bar{z}). \end{aligned}$$

This is a contradiction to the Pareto optimality of $(y, \bar{z})$.

2. First, let $\bar{z}_a = 0$. Then (9.4) directly yields that $y_i - y_j \leq s_a$. On the other hand, suppose that $\bar{z}_{\tilde{a}} = 1$ and $y_{\tilde{i}} - y_{\tilde{j}} \leq s_{\tilde{a}}$ for some $\tilde{a} = (\tilde{i}, \tilde{j}) \in \mathcal{A}_{change}$. Defining $\bar{z}'_{\tilde{a}} = 0$ and $\bar{z}'_a = \bar{z}_a$ for all other changing activities we obtain that $(y, \bar{z}')$ is feasible for (BDM) and

$$\begin{aligned} f_{\mathcal{E}}(y,\bar{z}) &= f_{\mathcal{E}}(y,\bar{z}') \\ f_{\mathcal{A}}(y,\bar{z}) &= \sum_{a \in \mathcal{A}_{change}} w_a \bar{z}_a \\ &> \sum_{a \in \mathcal{A}_{change}} w_a \bar{z}'_a = f_{\mathcal{A}}(y,\bar{z}'), \end{aligned}$$

since $w_{\tilde{a}} > 0$. Again, this is a contradiction to the Pareto optimality of $(y, \bar{z})$. □

Similar to the previous chapter, we can now define reduced solutions as follows.

Definition 9.3. *Let $(y, \bar{z})$ be a feasible solution of (BDM). Define*

$$\begin{aligned} y(\bar{z}) &= y(\mathcal{A}^{fix}(\bar{z})) \\ \bar{z}_a(y) &= \begin{cases} 0 & \text{if } y_i - y_j \le s_a \\ 1 & \text{otherwise} \end{cases} \end{aligned}$$

and the **reduced solution**

$$R^{BDM}(y, z) = (y^{red}, z(y^{red})),$$

where $y^{red} = y(\bar{z})$ is the time-minimal solution with respect to $\mathcal{A}^{fix}(\bar{z})$.

In contrast to (TDM) where we only knew that there always exists a reduced solution, Lemma 9.2 yields that **all** Pareto optimal solutions are reduced ones.

9.3 Lexicographic and Supported Efficient Solutions

Lexicographic Minimal Solutions

We first discuss the two lexicographic minimal solutions which can both be determined easily for (BDM). Minimizing in the order $(f_{\mathcal{A}}, f_{\mathcal{E}})$ we first have to minimize the weight of the missed connections. By setting $\bar{z}_a = 0$ for all $a \in \mathcal{A}_{change}$ we obtain $f_{\mathcal{A}} = 0$ which is the best possible solution value. To minimize $f_{\mathcal{E}}$ under this condition we determine $y = y(\mathcal{A}_{change})$ by Algorithm 12 and its objective function value $f_{\mathcal{E}}$ to obtain the lexicographic minimal solution $(y, \bar{z})$.

For the order $(f_{\mathcal{E}}, f_{\mathcal{A}})$ we proceed in two steps: First let $\mathcal{A}_{fix} = \emptyset$ (allowing all connections to be missed) and determine $y = y(\emptyset)$ again by Algorithm 12 to obtain a solution with minimal possible objective function value $f_{\mathcal{E}}$. To improve the corresponding value of $f_{\mathcal{A}}$ we calculate $\bar{z} = \bar{z}(y)$ according to Definition 9.3 above, i.e., we find out which connections are *really* missed in our particular solution y, yielding the lexicographic minimal solution $(y, \bar{z})$.

Supported Efficient Solutions

Supported efficient solutions are those solutions which can be determined by a scalarization of the two objectives, i.e., by solving the following minimization problem, see [Geo68], or Appendix B. Given $0 < \lambda < 1$ we have to solve

$$\min \lambda \sum_{i \in \mathcal{E}_{arr}} y_i + (1-\lambda) \sum_{a \in \mathcal{A}_{change}} w_a \bar{z}_a$$

such that

$$\begin{aligned} y_i &\geq d_i \text{ for all } i \in \mathcal{E} \\ y_i - y_j &\leq s_a \text{ for all } a = (i,j) \in \mathcal{A}_{wait} \cup \mathcal{A}_{drive} \\ -M\bar{z}_a + y_i - y_j &\leq s_a \text{ for all } a = (i,j) \in \mathcal{A}_{change} \\ y_i &\in \mathbb{N} \text{ for all } i \in \mathcal{E} \\ \bar{z}_a &\in \{0,1\} \text{ for all } a \in \mathcal{A}_{change}. \end{aligned}$$

Fortunately, this problem is equivalent to (TDM-const) (Section 8.3 on page 136).

Theorem 9.4. *Finding supported efficient solutions of (BDM) is equivalent to solving (TDM-const).*

Proof. Define $T = 1 - \lambda$, $w_i = \lambda$. (If T should be as large as it used to be, multiply all weights w_i and T by some large value.) □

This means we can use the results of Sections 8.4 and 8.3. Especially, the never-meet property gets important again, since it simplifies (BDM) significantly, if it is satisfied. Recall from Definition 8.18 that the delay management problem has the never-meet property, if for each $\mathcal{A}^{fix} \subseteq \mathcal{A}_{change}$ the time-minimal solution $y = y^0(\mathcal{A}^{fix})$ satisfies the following two conditions.
For all $j \in \mathcal{E}_{red-late}$:

1. If $(i_1, j), (i_2, j) \in \mathcal{A}^{fix} \cup \mathcal{A}_{wait} \cup \mathcal{A}_{drive}$, and $y_{i_2} > 0$ then $y_{i_1} = 0$.
2. If $(i_1, j) \in \mathcal{A}$, and $d_j > 0$ then $y_{i_1} = 0$.

As a consequence of Theorem 9.4 we hence get:

Corollary 9.5. *Let the delay management problem have the never-meet property. Then the supported efficient solution belonging to some fixed λ can be determined in time $O(|\mathcal{A}|)$.*

Proof. This follows from Theorem 9.4 and from the correctness of Algorithm 15 (see Theorem 8.34) on page 160. □

Without the never-meet property, but with zero slack times we can use Theorem 8.23 to obtain the next corollary.

Corollary 9.6. *Let $s_a = 0$ for all $a \in \mathcal{A}$. Then the supported efficient solution belonging to some fixed λ can be determined by linear programming.*

Proof. The corollary also follows from Theorem 9.4, this time in combination with Theorem 8.23 (see page 146). □

We remark that solving the scalarization in the general case can be done by Algorithm 14 (page 156), or by solving the relaxation using the lower bounds provided in Theorem 8.26 and Corollary 8.28 (see page 150).

9.4 Finding All Efficient Solutions

In this section we present an exact algorithm for finding all efficient solutions of (BDM). It is based on ideas from solving the *discrete time/cost tradeoff problem (DTCTP)* of project planning, and has been applied to delay management in [Gin01, GS02]. We start by describing (DTCTP).

The (DTCTP) in Project Planning

In project planning the classic goal is to determine the minimal project length. One possibility to finish a project quicker is to speed up some of the activities. Usually this is expensive, since new facilities have to be bought or more workers have to be hired. In this sense, we have two objective functions, namely the project length and the money to be spent, and the goal might be to

- minimize the project length with a given budget,
- attain some required project length as cheaply as possible, or
- find all efficient solutions for both objectives simultaneously.

If the cost-duration function is continuous for each activity, the problem has been widely studied, see [Elm77, Neu75] and references therein. In the discrete time/cost tradeoff problem (DTCTP), however, the duration $\bar{L}_a$ is a **discrete** non-increasing function g_a depending on the costs c. The possible cost-duration combinations of the respective activity are called *modes*, and are given by

$$\{(\bar{L}_a^m, c_a^m), m = 1, \ldots, M^a\}.$$

As in the continuous case, we have two objective functions in (DTCTP), namely

1. minimize the project length, and
2. minimize the costs.

In contrast to the continuous case, literature on (DTCTP) is rather sparse. The NP-hardness of the problem has been shown 1992 in [DDGW97]. Demeulemeester et al. [DHE96] gave two optimal procedures for that problem. The first algorithm is based on a procedure for finding the minimal number of reductions necessary to transform a general network to a series-parallel network, see [BKS92]. The second one minimizes the computational effort in enumerating alternative modes through a branch and bound tree. The solution method proposed in [DRF+98] uses a branch and bound procedure which computes lower bounds by making convex piecewise linear underestimations of the discrete time/cost curves of the activities. Those piecewise linear underestimations are used as input for an adapted version of the Fulkerson labeling algorithm for the linear time/cost trade-off problem.

For solving (BDM) we will adapt a solution method for (DTCTP) which has been suggested by [DHE96]. It is based on the following ideas.

Any project network $\bar{\mathcal{N}} = (\bar{\mathcal{E}}, \bar{\mathcal{A}})$ (according to Definition 7.3) can be reduced to one single arc by applying the following three operations (Figure 9.2), see [DHE96]:

Serial merge: Let a_1 be an arc from i to j, let a_2 be an arc from j to k, and let no other arc in $\bar{\mathcal{A}}$ be incident with node $j \in \bar{\mathcal{E}}$. Then merge a_1 and a_2 to one arc a going from i to k.

Parallel merge: Let $a_1, \ldots, a_p$ be p arcs going from i to j. Then merge all of them to one arc a going from i to j.

Node reduction: Let $i \in \bar{\mathcal{E}}$ such that only one incoming arc a_0 is incident with i, and let $a_1, \ldots, a_p$ be the outgoing arcs (or vice versa). Then merge a_0 with all of the arcs $a_1, \ldots, a_p$ serially to the new activities $a_1^0, \ldots, a_p^0$.

If no node reduction is needed in the reduction process, the network is called *series parallel*. An alternative approach to the critical path method (CPM) (see Section 7.2), is to successively shrink the network by applying the two merge operations and node reduction, until only one arc from s to t is left (see [Elm77]). The duration of the new arc a is calculated by

- $\bar{L}_a = \bar{L}_{a_1} + \bar{L}_{a_2}$ in a serial merge, and
- $\bar{L}_a = \max\{\bar{L}_{a_1}, \ldots, \bar{L}_{a_p}\}$ in a parallel merge.

Node reduction consists of p serial merges and is treated accordingly. We refer to Figure 9.2 for an illustration.

In the (DTCTP) we need to calculate the complete set of modes in each shrinking operation. In a serial or a parallel merge this can be done easily as follows.

Serial merge: $\{(\bar{L}_{a_1}^{m_1} + \bar{L}_{a_2}^{m_2}, c_{a_1}^{m_1} + c_{a_2}^{m_2}),$ for all modes m_1 of a_1, m_2 of $a_2\}$
Parallel merge: $\{(\max_{l=1,\ldots,p} \bar{L}_{a_l}^{m_l}, \sum_{l=1}^{p} c_{a_l}^{m_l}$ for all modes m_l of $a_l, l = 1, \ldots, p\}$

Note that many of the modes of the new arc a can be skipped (or, numerically better, need not be constructed) since they are dominated by other modes of a. Applying the two merging operations in a series-parallel network, one ends up with one single arc containing all efficient solutions (as its current set of modes). The (DTCTP) is hence easy solvable in series-parallel networks.

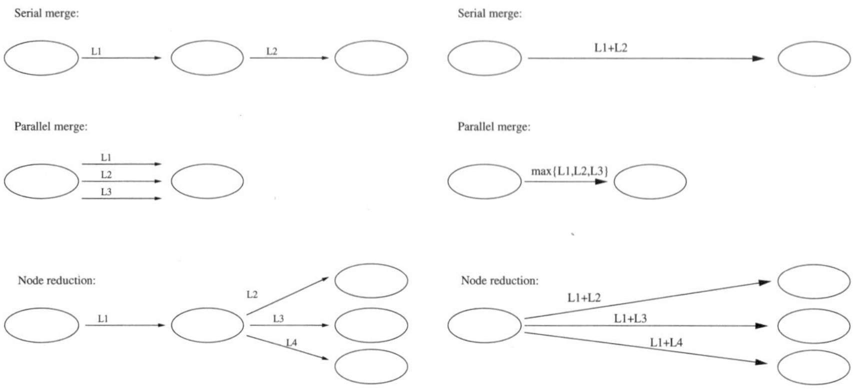

Fig. 9.2. Situation before and after a serial merge, parallel merge, or node reduction, when calculating the minimal completion time.

Unfortunately, node reduction cannot be performed as easily as the serial and the parallel merge in the bicriteria case, since one has to exclude the combination of two different modes of the same activity later on. Hence, for finding all efficient solutions one needs to fix the mode in a node reduction and to enumerate all possible combinations of modes. How to find a minimal set of activities to fix is described in [BKS92]. For more details, we refer to [DHE96].

To use these results for solving the bicriteria delay management problem, we will first discuss how the delay can be calculated correctly in the shrinking method in the next section and then show how (BDM) can be modeled and solved as (DTCTP).

Calculating the Delay Within the Shrinking Method

For answering the questions above we reconsider the delay management problem with fixed connections TT($\mathcal{A}^{fix}$) of Chapter 7. Recall that in this problem we have given a set of changing activities $\mathcal{A}^{fix} \subseteq \mathcal{A}_{change}$ for which we require

$$y_i - y_j \leq s_a$$

meaning that these connections must be maintained.

We now adapt the shrinking method sketched on page 182 to calculate the delay correctly in the delay management problem with fixed connections. Of course, this method yields the optimal solution by Theorem 7.5. But for the bicriteria delay management problem it is important to be able to calculate both objective functions in each intermediate step of the shrinking process.

To this end we introduce a second parameter $\bar{d}_a$ for each activity which calculates the delay, and, furthermore, we shrink the network in a specific order.

That is, we are only allowed to merge activities if no preceding activities exist any more. This is specified next. Note that during the reduction process parallel edges may occur, such that the notation $a = (i, j)$ in this section just means *activity a goes from event i to event j*. We now define the following network.

Notation 9.7. *Let $\mathcal{N} = (\mathcal{E}, \mathcal{A})$ and let $\mathcal{A}^{fix} \subseteq \mathcal{A}_{change}$. The corresponding* **delay network** $\mathcal{N}^d = (\mathcal{E}^d, \mathcal{A}^d)$ *is given through*

$$\mathcal{E}^d := \mathcal{E} \cup \{s, t\}$$
$$\mathcal{A}^d := \mathcal{A} \cup \{(s,i)^s : i \in \mathcal{E}\} \cup \{(s,i)^d : i \in \mathcal{E}_{del}\}$$
$$\cup \{(i,t) : i \textit{ is a maximal element of } \mathcal{E}\} \cup \{(s,t)\}$$
$$\bar{L}_a := \begin{cases} L_a & \textit{if } a \in \mathcal{A} \\ \Pi_i & \textit{if } a = (s,i)^s, i \in \mathcal{E} \\ \Pi_i + d_i & \textit{if } a = (s,i)^d, i \in \mathcal{E}_{del} \\ 0 & \textit{if } a = (i,t), i \in \mathcal{E} \\ T & \textit{if } a = (s,t). \end{cases}$$

The arcs $a = (s,i)^s$ are called **timetable arcs** *and the arcs $a = (s,i)^d$ are called* **delay arcs**.

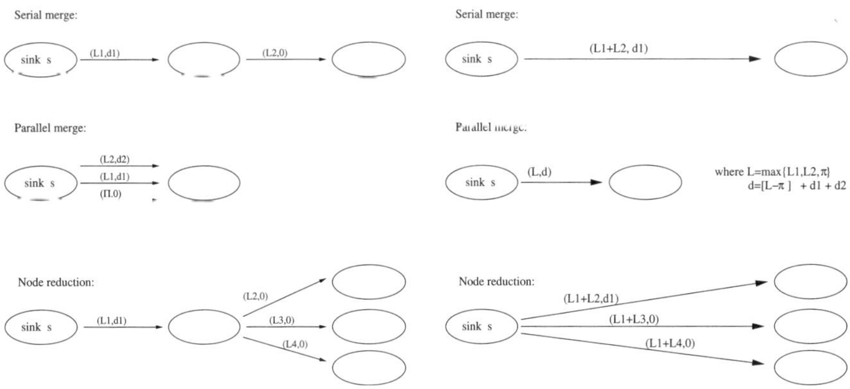

Fig. 9.3. Situation before and after a serial merge, parallel merge, or node reduction, when calculating the delay .

We remark that each node (except of the source s) is incident with exactly one incoming timetable arc.

We now come back to the delay management problem with fixed connections (TT($\mathcal{A}^{fix}$)) and present one more algorithm to solve it. The shrinking process of step 1 is depicted in Figure 9.3.

Algorithm 21: Shrinking method for solving TT($\mathcal{A}^{fix}$)

Input: $\mathcal{N}$, Π_i, d_i, L_a, $\mathcal{A}^{fix}$.
Output: Optimal (time-minimal) solution of TT($\mathcal{A}^{fix}$).
Step 0. Set $\bar{d}_a = 0$ for each arc, and let $\bar{L}_a$ be defined as in Notation 9.7.
Step 1. Apply the following operations if applicable:

Serial merge: Applicable for two activities a_1, a_2 if $a_1 = (s,i)$ $a_2 = (i,j)$ for $i,j \in \bar{\mathcal{E}}$, and no other activity is incident with i. Then delete i and replace a_1 and a_2 by one arc $a = (s,j)$ with

$$\bar{L}_a = \bar{L}_{a_1} + \bar{L}_{a_2} \tag{9.5}$$

$$\bar{d}_a = \bar{d}_{a_1} \tag{9.6}$$

Parallel merge: Applicable for activities $a_1, \ldots, a_p$ with $a_k = (s,i)$ for one common node $i \in \bar{\mathcal{E}}$, if there is no other incoming activity of event i. Let a^s be the (at most one) timetable arc. Then replace all a_k by one arc $a = (s,i)$ with

$$\bar{L}_a = \max\{\bar{L}_{a_1}, \ldots, \bar{L}_{a_p}\} \tag{9.7}$$

$$\bar{d}_a = \begin{cases} \sum_{a=a_1,\ldots,a_p} \bar{d}_a + \bar{L}_a - \bar{L}_{a^s} & \text{if } i \in \mathcal{E}_{arr} \\ \sum_{a=a_1,\ldots,a_p} \bar{d}_a & \text{if } i \notin \mathcal{E}_{arr}. \end{cases} \tag{9.8}$$

Node reduction: Applicable for activities $a_0, a_1, \ldots, a_p$ if $a_0 = (s,i)$ and $a_k = (i,j_k)$ with $i, j_k \in \bar{\mathcal{E}}$ for $k = 1, \ldots, p$, and no other activities are incident with i.
Then arbitrarily choose an outgoing arc of i, say a_1, delete i, and replace a_k, $k = 0, \ldots, p$ by p arcs $a_k^0 = (s, j_k)$, $k = 1, \ldots, p$.

$$\bar{L}_{a_k^0} = \bar{L}_{a_0} + \bar{L}_{a_k}, k = 1, \ldots, p \tag{9.9}$$

$$\bar{d}_{a_1^0} = \bar{d}_{a_0} \tag{9.10}$$

$$\bar{d}_{a_k^0} = 0, \quad k = 2, \ldots, p \tag{9.11}$$

Step 2 : If the network has been reduced to one single arc $a = (s,t)$ let $f_{\mathcal{E}} = \bar{d}_a$. Stop.

Before we prove its correctness, Algorithm 21 will be illustrated in the following example. Consider the situation shown in Figure 6.4 (see page 105), representing a PTN with two vehicles g and h that meet at station v_0, such that passengers can change between the two vehicles. The scheduled times and the lower bounds for each activity are given. Suppose that vehicle g arrives at v_0 with a delay of 10 minutes, i.e., $d_1 = 10$ and the duration of the corresponding delay arc $a = (s,1)^d$ is $\bar{L}_a = 18 + 10 = 28$. In Figure 9.4 the corresponding delay network, in which the delay arc $(s,1)$ has already been merged with its parallel timetable arc, is depicted.

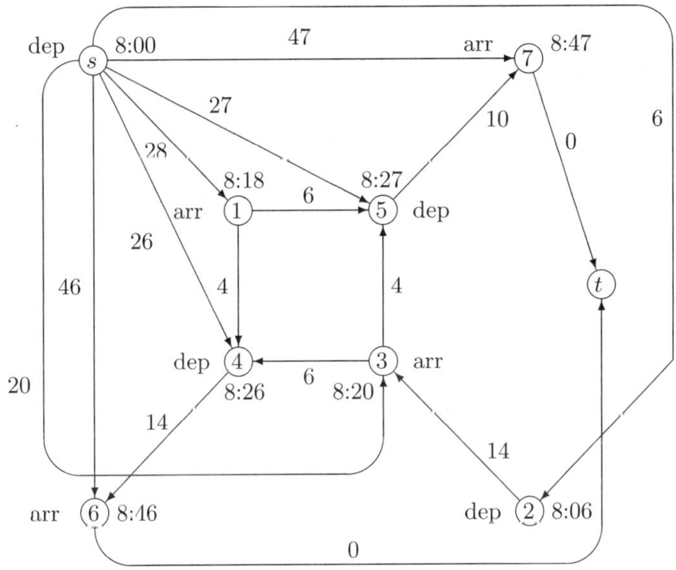

Fig. 9.4. $\mathcal{N}^d$ for the example. Event 1 is delayed by 10 minutes. The changing activities are dashed.

Table 9.1 gives the perturbed timetable x. In addition the original timetable Π and the delay for each event are given. The sum over all delays in the network equals 23 minutes, which is minimal under the assumption that all connections are kept.

Lemma 9.8. *$\bar{\mathcal{N}}$ can be reduced to one single arc from s to t by using serial merges, parallel merges, and node reduction as defined above.*

Proof. Let $\bar{\mathcal{N}}^c = (\bar{\mathcal{E}}^c, \bar{\mathcal{A}}^c)$ denote the reduced network with its set of nodes and arcs after c steps of reduction. Note that there exist directed paths from s to any of the nodes in $\bar{\mathcal{E}}^c$, and from each node in $\bar{\mathcal{E}}^c$ to t. Take a minimal node i in $\bar{\mathcal{E}}^c$ and let $a_1 = (s,i)$ be the arc from s to i i.e., there is no node $j \in \bar{\mathcal{E}}^c \setminus \{s\}$ with $j \prec i$.

i	event	x_i	Π_i	y_i
s		8:00	8:00	0
1	g, v_0, arr	8:28	8:18	10
2	h, v_3, dep	8:06	8:06	0
3	h, v_0, arr	8:20	8:20	0
4	g, v_0, dep	8:32	8:26	6
5	h, v_0, dep	8:34	8:27	7
6	g, v_2, arr	8:46	8:46	0
7	h, v_4, arr	8:47	8:47	0

$\sum_{i \in \mathcal{E}} (x_i - \Pi_i) = 23\,\text{min}$

Table 9.1. New timetable computed by Algorithm 21.

- If there are other arcs $a = (s, i)$, then a parallel merge is applicable.
- Otherwise, a_1 is the only incoming arc to event i. In this case, either $i = t$ and the reduction process is finished, or $i \neq t$ meaning that there exists at least one outgoing arc of node i, such that
  - a serial merge is applicable, or
  - a node reduction of node i can be performed. □

Lemma 9.9. *During the reduction process let $a = (s, i)$ be the only activity from s to i, $i \in \bar{\mathcal{E}}$. Then L_a contains the perturbed feasible timetable for event i of the optimal solution to TT($\mathcal{A}^{fix}$).*

Proof. Applying the shrinking operations together with the rules (9.5),(9.7), and (9.9) is equivalent to CPM (see [Elm77]). Furthermore, according to Theorem 7.5 we know that CPM yields an optimal solution to TT($\mathcal{A}^{fix}$). Together, the result follows. □

Lemma 9.10. *Let $a = (s, t)$ be the only remaining activity at the end of the shrinking process. Then $f_{\mathcal{E}} = \bar{d}_a$, i.e., $\bar{d}_a$ contains the sum of all arrival delays of the event-activity network $\mathcal{N}$.*

Proof. We show by induction that in each step of the shrinking process,

$$\sum_{a \in \bar{\mathcal{A}}} \bar{d}_a$$

contains the sum of all delays of arrival events in the set $\tilde{\mathcal{E}}$, consisting of

- all arrival events which have already been deleted during the process, and of
- *timetable events*, i.e., events $i \in \bar{\mathcal{E}}$ such that there exists exactly one incoming arc $a = (s, i)$.

After the initialization, all $\bar{d}_a = 0$. This is correct, since no activity has been deleted, and all delayed events $i \in \mathcal{E}_{del}$ have two incoming arcs $(s, i)^d$ and $(s, i)^s$, i.e., $\tilde{\mathcal{E}} = \emptyset$. Now we discuss each of the three shrinking operations.

Serial merge of an arc $a_1 = (s, i)$ and another arc $a_2 = (i, j)$ means to delete event i, which already has been a timetable event before. Since there exists at least one more arc to j (namely the timetable arc (s, j)) j will not enter $\tilde{E}$, hence $\tilde{E}$ does not change. On the other hand, activities a_1 and a_2 are replaced by a, but since $\bar{d}_{a_1} = \bar{d}_a$ and $\bar{d}_{a_2} = 0$ the sum of all $\bar{d}_{a'}$ over all $a' \in \bar{\mathcal{A}}^c$ does not change.

Node reduction is also correct, since it consists of d serial merges, where the delay of the timetable arc is transferred exactly to one of the new arcs.

Parallel merge of $a_1, \ldots, a_p$, all of them going from s to i means that i becomes a timetable event and is added to $\tilde{\mathcal{E}}$. We have to show that the new calculation of the delay now includes the delay of event i, if it is an arrival event. To this end, let a_1 be the unique timetable arc. Note that this means $L_{a_1} = \Pi_i$ contains the original timetable, while $\bar{L}_a = \max_{k=2,\ldots,p} \bar{L}_{a_k}$, contains the (perturbed) time for event i in an optimal solution of $(\mathrm{TT}(\mathcal{A}^{fix}))$ due to Lemma 9.10. Consequently, the delay of event i is zero, if $\bar{L}_a = \Pi_i$, otherwise the delay is given by $\bar{L}_a - \Pi_i$. Adding this new delay to the given ones $\bar{d}_{a_k}, k = 1, \ldots, p$ and using the induction hypothesis proves the result. In the case that i is not an arrival event, especially, $i = t$ no further delay needs to be added. Replacing $a_1, \ldots, a_p$ by one arc a is correct since all the delays are added to the delay of the new arc a. □

Lemma 9.9 and 9.10 together finally show the desired correctness of our procedure.

Theorem 9.11. *Algorithm 21 yields an optimal solution to TT($\mathcal{A}^{fix}$).*

The Algorithm for (BDM)

To find the set of all efficient solutions of (BDM) we will now transfer the algorithm by [DHE96] for the discrete time/cost trade-off problem (DTCTP) to the bicriteria delay management problem.

First, we have to define modes for each arc as follows. For all activities $a \in \mathcal{A}_{change}$ we define two modes, given by $(L_a, 0)$ and $(-\infty, w_a)$, representing that we either maintain the connection, i.e., the duration of the changing arc has to be included in the calculation, or we do not maintain the connection which means that we lose a weight of w_a customers. All other activities (waiting, driving) as well as the timetable arcs in the construction of the delay network $\bar{\mathcal{N}}^d$ only get one single mode $(\bar{L}_a, 0)$. A third parameter, calculating the delay in each step is also necessary. The modes are given by the triple (duration, delay, weight of lost connection). A mode m_1 dominates m_2, if m_2 is dominated in the two last components of the triples.

Algorithm 22: Finding all efficient solutions of (BDM)

Input: $\mathcal{N}$, d_i, Π_i, L_a, w_a.
Output: All efficient solutions of (BDM).
0. Initialize: Set $\bar{d}_a = 0$ for each arc, and let $\bar{L}_a$ be defined as in Notation (9.7). Initialize one mode $m = (\bar{L}_a, \bar{d}_a, 0)$ for all $a \notin \mathcal{A}_{change}$ and two modes $m_1^a = (L_a, \bar{d}_a, 0), m_2^a = (-\infty, d_a, w_a)$ for all $a \in \mathcal{A}_{change}$.
Step 1: Apply the following operations if applicable:
 1.1. Serial merge: Applicable for two activities a_1, a_2 if $a_1 = (s, i)$, $a_2 = (i, j)$ for $i, j \in \bar{\mathcal{E}}$, and no other activity is incident with i. Then delete i and calculate the modes of the new activity a by combining each possible combination of modes, i.e.,

$$\{(\bar{L}_{a_1}^{m_1} + \bar{L}_{a_2}^{m_2}, \bar{d}_{a_1}^{m_1}, \bar{w}_{a_1}^{m_1} + \bar{w}_{a_2}^{m_2}), \text{ for all modes } m_1 \text{ of } a_1, m_2 \text{ of } a_2\}$$

 and deleting dominated ones.
 1.2. Parallel merge: Applicable for activities $a_1, \ldots, a_p$ with $a_k = (s, i)$ for one common node $i \in \bar{\mathcal{E}}$, if there is no other incoming activity of event i.
 Let a^s be the (at most one) timetable arc. Then the modes of the new activity a are given by

$$\{(\max_{l=1}^{p} \bar{L}_{a_l}^{m_l}, \sum_{l=1}^{p} \bar{d}_{a_l}^{m_l} + \max_{l=1}^{p} \bar{L}_{a_l}^{m_l} - \bar{L}_{a^s}, \sum_{l=1}^{p} \bar{w}_{a_l}^{m_l})$$

$$\text{for all modes } m_l \text{ of } a_l, l = 1, \ldots, p$$

 in the case that $i \in \mathcal{E}_{arr}$. For $i \notin \mathcal{E}_{arr}$ the modes are given by

$$\{(\max_{l=1}^{p} \bar{L}_{a_l}^{m_l}, \sum_{l=1}^{p} \bar{d}_{a_l}^{m_l}, \sum_{l=1}^{p} \bar{w}_{a_l}^{m_l})$$

$$\text{for all modes } m_l \text{ of } a_l, l = 1, \ldots, p\},$$

 where dominated modes are deleted.
 1.3. Node reduction: Applicable for activities $a_0, a_1, \ldots, a_p$ if $a_0 = (s, i)$ only has one single mode, $a_k = (i, j_k)$ with $i, j_k \in \bar{\mathcal{E}}$ for $k = 1, \ldots, p$, and no other activities are incident with i. Then arbitrarily choose an outgoing arc of i, say a_1, delete i, and calculate the modes of the new activity $a_k^0, k = 2, \ldots, p$ by

$$\{(\bar{L}_{a_0} + \bar{L}_{a_k}^{m}, 0, \bar{w}_{a_k}^{m}) \text{ for all modes } m \text{ of } a_k\}$$

 and of a_1^0 by

$$\{(\bar{L}_{a_0} + \bar{L}_{a_1}^{m}, d_{a_0}, \bar{w}_{a_0} + \bar{w}_{a_1}^{m}) \text{ for all modes } m \text{ of } a_k\}.$$

Step 2.
 2.1 Final reduction If the network has been reduced to one single arc goto 3.

```
    2.2 Fixing an activity Otherwise, choose activity a = (s,i) such that
        i is the earliest node still in the network, fix one mode of
        activity a and perform node reduction as in step 1.3.
        Goto 1.
Step 3: Add the modes of the new solution, delete all non-efficient ones
    (compared to other solutions obtained in previous steps) and goto 1,
    fixing another combination of modes during step 2.2
    If all combinations of modes have been calculated, stop.
```

Theorem 9.12. *Algorithm 2 finds all efficient solutions of (BDM).*

Proof. Using the result of Theorem 9.10 we know that the delay $f_{\mathcal{E}}$ has been calculated correctly when reaching step 2.1. It is also clear that the final value of $\bar{w}$ in the remaining activity from s to t equals $f_{\mathcal{A}}$. If all solutions had been determined for each possible combination of modes, the resulting set would contain all efficient solutions. Since a solution can never be Pareto, if parts of it are dominated (i.e., can be replaced by a better solution) it is feasible to delete dominated modes for single activities during the reduction process. Since in the final step, all remaining dominated solutions are deleted, we end up with the set of non-dominated solutions in the end. □

Consider again our example on page 186. Figure 9.5 shows the corresponding DTCTP. First a serial merge operation can be applied to activities 6 and 10. Denote the new activity by 17 with mode (20,0,0) and merge it in parallel with activity 5 which yields the new activity 18 with mode (20,0,0). Now a node reduction step is performed for event 1, adding the costs of activity 1 to a succeeding activity, e.g., activity 8 and thus obtain activity 19 with mode (32,10,0). Activity 9 can then be merged serially with activity 1, this time without adding the costs to avoid double counting. This yields activity 20 with modes (34,0,0) and ($-\infty$,0,1). Now merge activity 3 and 19 in parallel. Since activity 3 is a timetable arc and the duration of activity 19 is greater than the one of activity 3, the difference of 6 minutes has to be added to the delay of the new activity, which is numbered by 21 and has mode (32,16,0). Another parallel merge operation can be performed with activities 2 and 20 which results in activity 22 with modes (34,7,0),(27,0,1). The delay of 7 minutes in the first mode results from the difference in the duration of activities 20 and 2 ($34 - 27 = 7$). Figure 9.6 shows the network after these reduction steps. Then it can be continued in the same way by performing a node reduction with event 3. Table 9.2 shows the complete reduction plan.
In step 3 of the algorithm we compare the obtained solutions with solutions found in previous iterations. Since in this example there were no activities to fix with more than one mode, the computation can be stopped and yields two efficient solutions: the first one with no missed connections and 23 minutes of

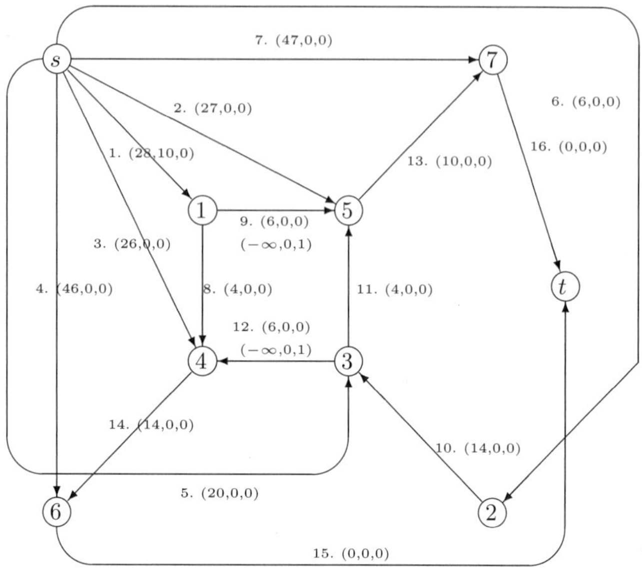

Fig. 9.5. $\bar{\mathcal{N}}^d$ with the modes for DTCTP.

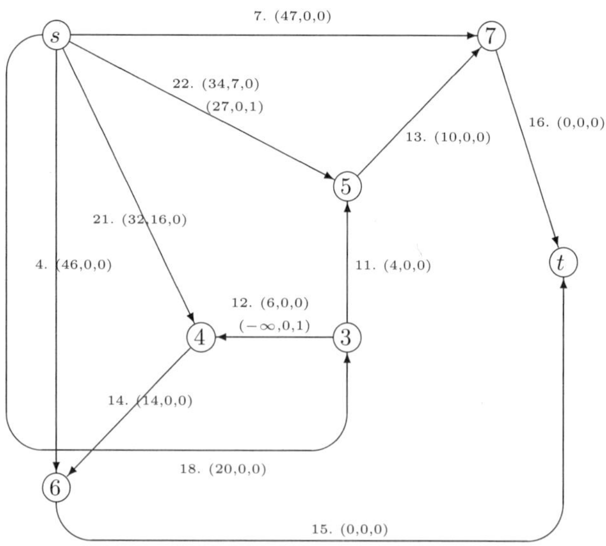

Fig. 9.6. Network after the first six reduction steps.

Action	Activities	New Modes
1.SERIES	[6,10]	17 (20,0,0)
2.PARALLEL	[5,17]	18 (20,0,0)
3.REDUCE	[1,8]	19 (32,10,0)
4.REDUCE	[1,9]	20 (34,0,0),($-\infty$,0,1)
5.PARALLEL	[3,19]	21 (32,16,0)
6.PARALLEL	[2,20]	22 (34,7,0),(27,0,1)
7.REDUCE	[18,11]	23 (24,0,0)
8.REDUCE	[18,12]	24 (26,0,0),($-\infty$,0,1)
9.PARALLEL	[21,24]	25 (32,16,0)
10.PARALLEL	[22,23]	26 (34,7,0),(27,0,1)
11.SERIES	[25,14]	27 (46,16,0)
12.PARALLEL	[4,27]	28 (46,16,0)
13.SERIES	[26,13]	29 (44,7,0),(37,0,1)
14.PARALLEL	[7,29]	30 (47,7,0),(47,0,1)
15.SERIES	[28,15]	31 (46,16,0)
16.SERIES	[30,16]	32 (47,7,0),(47,0,1)
17.PARALLEL	[31,32]	33 (47,23,0),(47,16,1)
18.EVALUATE	33	

Table 9.2. Reduction plan for the example.

total delay and the second one with one missed connection and 16 minutes of total delay.

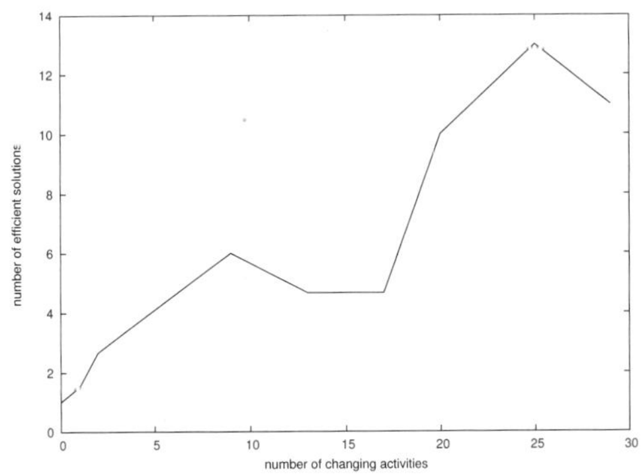

Fig. 9.7. Number of efficient solutions as function of the size of $\mathcal{A}_{change}$.

Instead of fixing all possible combinations of modes, a branch and bound approach can be applied to make the procedure more efficient. Nevertheless, our numerical experiments indicate that Algorithm 22 runs rather quickly if the relevant time interval to be considered is not too large. Evaluating the delay and the missed connections for all events and activities within the next

60 minutes could in most cases be done within a running time of less than a minute on a standard personal computer. Both the running time and the number of efficient solutions depend on the number of changing activities that need to be considered. This behavior is depicted in Figure 9.7, details can be found in [GS02, Gin01].

The nice behavior of Algorithm 22 on our practical data is due to the fact that the number of changing activities outside of the municipal areas is relatively small, such that the event-activity network is close to a series-parallel one. Summarizing, the method seems to have the potential to be used as an online decision support procedure.

10

Extensions

In the first part of this chapter we present a very general model for the delay management problem, called *general delay management problem* (GDM). It contains all models discussed so far as special cases and allows us to define many other interesting objective functions. The goal in the general delay management problem is to minimize the following two objectives simultaneously:

- the number of customers missing a connection and
- the amount of the additional delay of the remaining customers,

where we allow two different subsets $\mathcal{P}_1, \mathcal{P}_2 \subseteq \mathcal{P}$, specifying the paths relevant for either objective.

In the second part we discuss some requirements arising in practice that are currently under research. In particular in railway delay management problems, additional requirements modeling the limited capacity of the tracks need to be taken into account.

Chapter 10 is structured as follows: We discuss the general delay management problem (GDM), give an integer programming formulation and point out its relation to the models (TT), (TDM), and (BDM) discussed before. Furthermore we briefly mention requirements for delay management which appear in practice and first solution approaches to including them in our models.

10.1 The General Delay Management Problem

Here we present a very general model for the delay management problem. We propose a bicriteria formulation, which takes into consideration that arriving with a delay and missing a connection may have different utility functions. The specification, *how* more important it is to miss a connection, is left open since experience tells us that this decision strongly depends on the preferences of the respective planner of the public transportation company.

To formalize the objective functions of (GDM), let two sets of customers' paths $\mathcal{P}_1, \mathcal{P}_2$ be given and let $\mathcal{P} = \mathcal{P}_1 \cup \mathcal{P}_2$. (Paths have been defined in Chapter 8.) Then consider passengers traveling along a path $p \in \mathcal{P}$ through the public transportation network. We are interested in the following two cases.

Case 1: If $p \in \mathcal{P}_1$ and all connections on path p are maintained, then we add the delay of a passenger using path p to our first objective. This delay of a passenger using path p equals the arrival delay of his last vehicle $g(p)$ at his destination station $v(p)$, given by

$$x_{arr}{}^{v(p)}_{g(p)} - \Pi_{arr}{}^{v(p)}_{g(p)}.$$

Case 2: If $p \in \mathcal{P}_2$ and at least one connection on p is missed, then we count all passengers using path p in the second objective.

The general delay management problem can now formally be stated as follows.

(GDM)

Given PTN, $F, \mathcal{U}$, *minimal necessary times for driving, waiting, and changing, a feasible timetable* Π_{arr}, Π_{dep}, *a set of paths* $\mathcal{P}$ *through* PTN *with weights* w_p *for all* $p \in \mathcal{P}$, *and a set of delayed events* $\mathcal{E}_{del}$, *find a perturbed feasible timetable* x_{arr}, x_{dep}, *such that both*

$$f_1 = \sum_{p \in \mathcal{P}_1 : p \text{ is maintained}} w_p \left(x_{arr}{}^{v(p)}_{g(p)} - \Pi_{arr}{}^{v(p)}_{g(p)} \right) \quad \text{and}$$

$$f_2 = \sum_{p \in \mathcal{P}_2 : p \text{ is missed}} w_p,$$

are minimized.

By defining the sets $\mathcal{P}_1, \mathcal{P}_2$ appropriately, one can investigate many interesting objective functions as special cases of f_1 and f_2. For example, defining $\mathcal{P}_1$ as the set of paths used by the drivers of the PTN, f_1 will give the sum of delays of the drivers when coming back to the depot at the end of their shifts, and is hence the sum of the overtime over all drivers.

Integer Programming Formulation

(GDM) can be formulated as the following integer program.

(GDM)

$$\min \begin{pmatrix} \sum_{p\in\mathcal{P}_1} w_p y_{i(p)}(1-z_p) \\ \sum_{p\in\mathcal{P}_2} w_p z_p \end{pmatrix},$$

such that

$$y_i \geq d_i \quad \forall i \in \mathcal{E}_{del} \tag{10.1}$$

$$y_i - y_j \leq s_a \quad \forall a=(i,j) \in \mathcal{A}_{wait} \cup \mathcal{A}_{drive} \tag{10.2}$$

$$-M\bar{z}_a + y_i - y_j \leq s_a \quad \forall a=(i,j) \in \mathcal{A}_{change} \tag{10.3}$$

$$M\bar{z}_a - y_i + y_j \leq M - \frac{1}{2} - s_a \quad \forall a=(i,j) \in \mathcal{A}_{change} \tag{10.4}$$

$$z_p - \sum_{a\in\mathcal{A}_{change}\cap p} \bar{z}_a \leq 0 \quad \forall p \in \mathcal{P} \tag{10.5}$$

$$z_p - \bar{z}_a \geq 0 \quad \forall p \in \mathcal{P}, a \in \mathcal{A}_{change}\cap p \tag{10.6}$$

$$y_i \leq T \quad \forall i \in \mathcal{E} \tag{10.7}$$

$$y_i \in \mathbb{N} \quad \forall i \in \mathcal{E}$$

$$z_p \in \{0,1\} \quad \forall p \in \mathcal{P}$$

$$\bar{z}_a \in \{0,1\} \quad \forall p \in \mathcal{P}.$$

The following observations make sure that the integer programming formulation is a correct model for (GDM).

Lemma 10.1. *In each feasible solution of (GDM) the following holds.*

1. $\bar{z}_a = 0$ *if and only if connection* a *is maintained.*
2. $z_p = 0$ *if and only if all connections on path* p *are maintained.*

Proof. 1. First consider a connection $a \in \mathcal{A}_{change}$.

$\Longrightarrow$: Let $\bar{z}_a = 0$. Constraint (10.3) then directly gives $y_i - y_j \leq s_a$, meaning that this connection is maintained according to Definition 6.3.

$\Longleftarrow$: Let a be a maintained connection, i.e., $y_i - y_j \leq s_a$. Then constraint (10.4), together with the integrality of the variables, forces $\bar{z}_a = 0$.

2. Now consider some path $p \in \mathcal{P}$.

$\Longleftarrow$: Let $z_p = 0$. Then constraint (10.6) yields $z_a = 0$ for all $a \in \mathcal{A}_{change} \cap p$, meaning that all these connections are maintained due to part 1 of Lemma 10.1.

$\Longrightarrow$: Let a be maintained for all $a \in p \cap \mathcal{U}$. Again, using the first part of Lemma 10.1 we obtain that $\bar{z}_a = 0$ for all $a \in p \cap \mathcal{A}_{change}$. Hence, $\sum_{a\in\mathcal{A}_{change}\cap p} \bar{z}_a = 0$ and constraint (10.5) yields $z_p = 0$. □

We also have to clarify the size of M.

Lemma 10.2. *Choosing* $M \geq T + s_a + \frac{1}{2}$ *for all* $a \in \mathcal{A}_{change}$ *is large enough for constraints (10.4), while for constraint (10.3)* $M \geq y_i - y_j - s_a$ *for all* $a = (i,j) \in \mathcal{A}_{change}$ *is sufficient.*

Relations to the Models Discussed Before

Relation to (TT)

f_{TT} can be interpreted as a special case of f_1 in (GDM), namely, if the set of paths $\mathcal{P}_1$ only contains paths p such that for all $a \in p$: $a \in \mathcal{A}^{fix}$. Defining

$$\begin{aligned}\mathcal{P}^{fix} &= \{p \in \mathcal{P} : \text{ for all } a \in p : a \in \mathcal{A}^{fix}\} \text{ and} \\ w_i^{fix} &= \sum_{p \in \mathcal{P}^{fix} : i(p)=i} w_p\end{aligned}$$

we conclude that

$$f_1 = \sum_{p \in \mathcal{P}^{fix}} w_p y_{i(p)} = \sum_{p \in \mathcal{P}_1 : z_p = 0} w_i^{fix} y_i = f_{rmTT}.$$

Relation to (TDM)

Here we prove that (TDM) is a scalarization of (GDM), and hence all solutions of (TDM) are (supported) efficient solutions of the general model (GDM). To this end, we use the formulation (TDM-A) on page 122.

Theorem 10.3. *Let $d_i \leq D < T$ for all $i \in \mathcal{E}_{del}$ and let (y, z) be an optimal solution of (TDM-A). Then there exists $\bar{z}$ such that $(y, z, \bar{z})$ is a (supported) efficient solution of (GDM-B).*

Proof. We show that (TDM-A) is equivalent to the scalarization of (GDM), given by

$$\min f_1(y, \bar{z}, z) + T f_2(y, \bar{z}, z)$$

such that (10.1) – (10.7) and the integrality constraints are satisfied (see page 197), then the result follows from weighted sum scalarization with $\lambda_1 = \frac{1}{1+T}$, $\lambda_2 = \frac{T}{1+T}$ due to [Geo68], or see Appendix B.

(TDM-A) $\Longrightarrow$ (GDM): Consider a feasible solution (y, z) of (TDM-A) with objective function value f. We show that

$$(y^{red}, z^{red}, \bar{z}),$$

with

$$\begin{aligned}y^{red} &= y(\mathcal{A}^{fix}(z)), \\ z^{red} &= z(y^{red}), \text{ and} \\ \bar{z}_a &= \begin{cases} 0 & \text{if } y_i^{red} - y_j^{red} \leq s_a \\ 1 & \text{otherwise} \end{cases}\end{aligned}$$

is a feasible solution of (GDM) with the same or a better objective function value. From Lemma 8.3 we already know that $R^A(y, z) = (y^{red}, z^{red})$ is feasible for (TDM-A) and has an equal or better objective value; hence it only remains to show that $(y^{red}, z^{red}, \bar{z})$ satisfies constraints (10.3) – (10.7) of (GDM), i.e.,

$$\begin{array}{ll}
(10.3) & -M\bar{z}_a + y_i^{red} - y_j^{red} \le s_a \;\; \forall a = (i,j) \in \mathcal{A}_{change} \\
(10.4) & M\bar{z}_a - y_i^{red} + y_j^{red} \le M - \frac{1}{2} - s_a \;\; \forall a = (i,j) \in \mathcal{A}_{change} \\
(10.5) & z_p^{red} - \sum_{a \in \mathcal{A}_{change} \cap p} \bar{z}_a \le 0 \;\; \forall p \in \mathcal{P} \\
(10.6) & -z_p^{red} \le -\bar{z}_a \;\; \forall p \in \mathcal{P}, a \in \mathcal{A}_{change} \cap p \\
(10.7) & y_i \le T \;\; \forall i \in \mathcal{E}.
\end{array}$$

(10.3): From the definition of $\bar{z}$ we know that $y_i^{red} - y_j^{red} \le s_a$ if $\bar{z}_a = 0$; and for $\bar{z}_a = 1$ there is nothing to show.

(10.4): This is trivially satisfied for $\bar{z}_a = 0$. On the other hand, if $\bar{z}_a = 1$, we get (due to the integrality of y^{red})

$$\begin{aligned}
& y_i^{red} - y_j^{red} > s_a \\
\Longrightarrow\; & y_i^{red} - y_j^{red} \ge s_a + \frac{1}{2}.
\end{aligned}$$

(10.5): For $z_p^{red} = 1$ we know that there exists some $a = (i,j) \in p$ such that $y_i^{red} - y_j^{red} > s_a$. Consequently, $\bar{z}_a = 1$ and hence

$$z_p^{red} \le \sum_{a \in \mathcal{A}_{change} \cap p} \bar{z}_a.$$

The case $z_p^{red} = 0$ is trivial.

(10.6): Let $a \in p$. The case $z_p^{red} = 1$ is trivial. If $z_p^{red} = 0$ then we know that for all $a = (i,j) \in p$: $y_i^{red} - y_j^{red} \le s_a$ from (8.3), and hence $\bar{z}_a = 0$, which shows (10.6).

(10.7): From Corollary 7.9 we know that $y_i^{red} \le D$ for all $i \in \mathcal{E}$, if $d_i \le D$ for all $i \in \mathcal{E}_{del}$; hence $y_i^{red} \le T$ for all $i \in \mathcal{E}$.

(GDM) $\Longrightarrow$ (TDM-A): Since there are only some constraints missing in the formulation of (TDM-A), each feasible solution $(y, \bar{z}, z)$ of (GDM) yields a feasible solution (y, z) of (TDM-A). Furthermore, for any feasible solution $(y, \bar{z}, z)$ of (GDM) we have that both objective functions f_1, f_2 do not depend on $\bar{z}$, hence

$$f_1(y, \bar{z}, z) + T f_2(y, \bar{z}, z) = f_{TDM-A}(y, z)$$

and this direction is also satisfied. □

Relation to (BDM)

Finally, we remark that the bicriteria delay management problem (BDM) (see page 175) is also a special case of (GDM). This is stated more precisely in the following lemma.

Lemma 10.4. *(BDM) is equivalent to (GDM) in the case that*

$$\mathcal{P}_1 = \{p_a : a \in \mathcal{A}_{drive}\}$$
$$\mathcal{P}_2 = \{p_a : a \in \mathcal{A}_{change}\}$$
$$w_p = \begin{cases} 1 & \textit{for all } p \in \mathcal{P}_1 \\ w_a & \textit{for all } p = p_a \in \mathcal{P}_2, \end{cases}$$

where for $a = (i,j) \in \mathcal{A}$ the corresponding path p_a is simply given by a sequence of the two events, i.e., $p_a = (i,j)$.

Note that to represent the paths of $\mathcal{P}_2$ in PTN, we have to add two more events to all paths $\bar{p}_a$, namely

$$\bar{p}_a = (i', i, j, j')$$

where $(i', i) \in \mathcal{A}_{drive}$ and $(j, j') \in \mathcal{A}_{drive}$ are the unique driving activities leading into event i, and going out of event j, respectively.

Proof. (GDM) $\Longrightarrow$ (BDM): Let $(y, \bar{z}, z)$ be a feasible solution of (GDM). Then $(y, \bar{z})$ is clearly feasible for (BDM). Furthermore, note that $(y, \bar{z}, z)$ satisfies constraints (10.5) and (10.6) given by

$$\text{(10.5)} \quad z_p - \textstyle\sum_{a \in \mathcal{A}_{change} \cap p} \bar{z}_a \leq 0 \quad \forall p \in \mathcal{P}$$
$$\text{(10.6)} \quad z_p \geq \bar{z}_a \quad \forall p \in \mathcal{P}, a \in \mathcal{A}_{change} \cap p.$$

Since no path of $\mathcal{P}_1$ contains a changing activity we conclude from (10.5) that

$$z_p = 0.$$

For $\bar{p}_a \in \mathcal{P}_2$ we know that a is the only changing activity contained in $\bar{p}_a$, hence, (10.5) and (10.6) together yield

$$z_{\bar{p}_a} = \bar{z}_a.$$

For the objective functions we hence obtain

$$\begin{aligned} f_{\mathcal{E}}(y, \bar{z}) &= \sum_{i \in \mathcal{E}_{arr}} y_i \\ &= \sum_{p \in \mathcal{P}_1} w_p y_{i(p)} \\ &= \sum_{p \in \mathcal{P}_1} w_p y_{i(p)} (1 - z_p) \\ &= f_1(y, \bar{z}, z), \end{aligned}$$

and,

$$\begin{aligned} f_{\mathcal{A}}(y, \bar{z}) &= \sum_{a \in \mathcal{A}_{change}} w_a \bar{z}_a \\ &= \sum_{a \in \mathcal{A}_{change}} w_{\bar{p}_a} z_{\bar{p}_a} \\ &= \sum_{p \in \mathcal{P}_2} w_p z_p \\ &= f_2(y, \bar{z}, z). \end{aligned}$$

(BDM) $\Longrightarrow$ (GDM): Now let $(y, \bar{z})$ be a Pareto solution of (BDM). Define

$$z_p = \begin{cases} 0 & \text{if } p \in \mathcal{P}_1 \\ \bar{z}_a & \text{if } p = \bar{p}_a \in \mathcal{P}_2. \end{cases}$$

We have to show that $(y, \bar{z})$ satisfies all constraints of (GDM) and leads to the same objective function values for both objectives. Constraints (10.1),(10.2), and (10.3) are identical to the first three constraints in (BDM), and constraints (10.5) and (10.6) hold due to the definition of z_p. According to part 1 of Lemma 9.2 we know that $y = y(\mathcal{A}^{fix}(\bar{z})$, and hence $y_i \leq \max_{i \in \mathcal{E}} d_i \leq T$, yielding (10.7).

Finally, consider (10.4) given by

$$(10.4) \quad M\bar{z}_a - y_i + y_j \leq M - \frac{1}{2} - s_a \quad \forall a = (i, j) \in \mathcal{A}_{change}.$$

For $z_a = 0$ this is trivially satisfied. On the other hand, from part 2 of Lemma 9.2 we know that for $\bar{z}_a = 1$ we have that $y_i - y_j > s_a$, hence

$$-y_i + y_j \leq -s_a - 1 \leq -s_a - \frac{1}{2},$$

establishing (10.4). Analogously to the first part of this proof, both objective functions coincide. □

This yields the following corollary.

Corollary 10.5. *The general delay management problem (GDM) is NP-hard, even if all slack times are zero, $\mathcal{P}_1 = \mathcal{P}_2$, and no two connections are contained in the same connected component of* PTN.

10.2 Railway and Bus Specific Requirements

In practice, there are other effects which need to be considered when dealing with the delay management problem. Many of them occur since passing and overtaking of two trains can not be done anywhere on the tracks. E.g., on a single-track line, two trains have to pass each other in a station, and also overtaking is only possible in a set of certain locations. Finding train schedules on single-track lines has been considered, e.g., in [XC94, HKF96, Hig97], and the estimation of delays for single-track train traffic is discussed, e.g., in [CH90]. As already mentioned in the introduction, a recent overview about scheduling and re-scheduling of trains is given in [Tör05a].

In the following we briefly outline how railway capacity requirements can be included in the delay management problem.

Oncoming traffic: On single track lines the oncoming traffic has to be taken into consideration, i.e., a vehicle is only allowed to leave if the oncoming vehicle has arrived. This effect can be modeled by introducing some new connections (g, h, v) representing that vehicle h is not allowed to depart before the oncoming train has arrived. Fixing $\bar{z}_{ghv} = 0$ it is possible to ensure that all these new connections are maintained, if the sequence of the trains is fixed, see Figure 10.1.

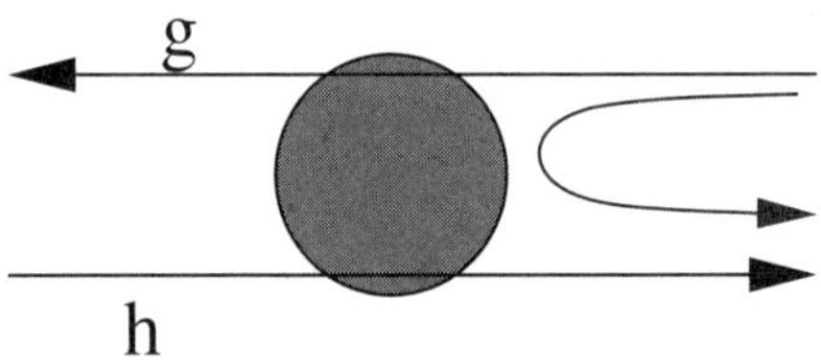

Fig. 10.1. Vehicle h is only allowed to depart if vehicle g has arrived.

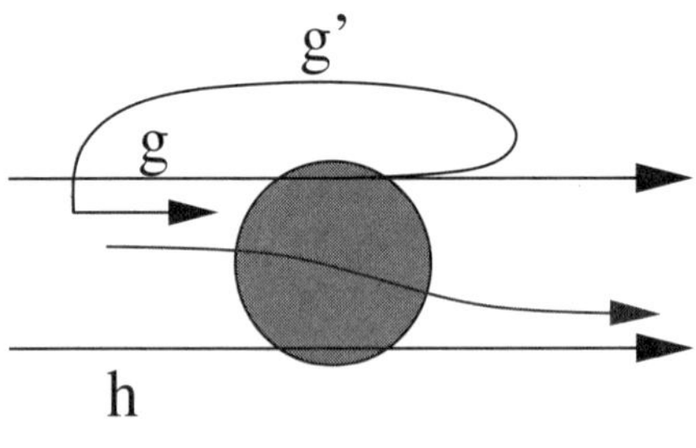

Fig. 10.2. Vehicle h is only allowed to leave if (the fast) vehicle g has overtaken.

Overtaking: Analogously, overtaking is a problem on single-track railways, since it can only be done at a few points on the tracks in this case. This means that a (slow) vehicle h is only allowed to leave such a point, if the (fast) vehicle g has overtaken, i.e., if it has already departed. If the point where overtaking should take place is known beforehand, this requirement can also be modeled as an artificial activity in the event-activity network $\mathcal{N}$, namely from (g, k, dep) to (h, v, dep). Note that this is neither a changing, nor a waiting, or driving activity since it connects two departure events. In PTN we can illustrate the situation as in Figure 10.2.

Safety requirements: Finally, we discuss safety requirements. These have to make sure that two vehicles do not depart at the same time but with at least a distance of s minutes. Again, if we know the sequence of the departures beforehand, say, g departs before h, then we add the same type of activity

$$a = ((g, v, \text{dep}), (h, v, \text{dep}))$$

as for overtaking, but with a minimal duration $L_{ghv} = s$. Fixing $\bar{z}_{ghv} = 0$ makes sure that the safety requirement will be satisfied in all perturbed timetables.

In the (more likely) case that the sequence of trains or the places for overtaking are not known, we end up with disjunctive constraints, which make the problem much harder to solve. Fortunately, we can still obtain an integer programming formulation by using additional binary variables to determine the sequence within the integer program. These are given (exemplary for the oncoming traffic, if passing is only possible at stations in V) as follows.

$$s_{ghv} = \begin{cases} 1 & \text{if vehicle } g \text{ is at station } v \text{ before vehicle } h \\ 0 & \text{otherwise.} \end{cases}$$

To ensure that at least one of these constraints is satisfied, we add constraints like

$$\sum_{v \in V} s_{ghv} \geq 1.$$

An exact formulation and solution procedures for this extension are currently under research. In the the graph-theoretic approach, the idea is to first construct a changing arc for each possible sequence of trains, or for each possible point suitable for overtaking, and then to make sure that at least one of these arcs is maintained. This leads to disjunctive constraints, and might be solvable using the concept of alternative graphs.

Note that modeling such capacity constraints exactly, a microscopic view is needed instead of the macroscopic model used so far. This includes modeling many *blocks* instead of just one single edge between two stations, and looking at platforms instead of stations. Integrated models combining the wait/depart decision and the train re-scheduling aspects are currently under research within a project together with *Deutsche Bahn*, see [BGJ+05].

In bus transportation the above constraints are not needed, but in this case, vehicle schedules and driver schedules may have an effect on the model, since a delayed vehicle or a delayed driver cannot start his next piece of duty in time. These effects can easily be modeled by introducing new connections, which have to be maintained to make sure that a delay is taken over to the next piece of duty.

Part III

Tariff Planning

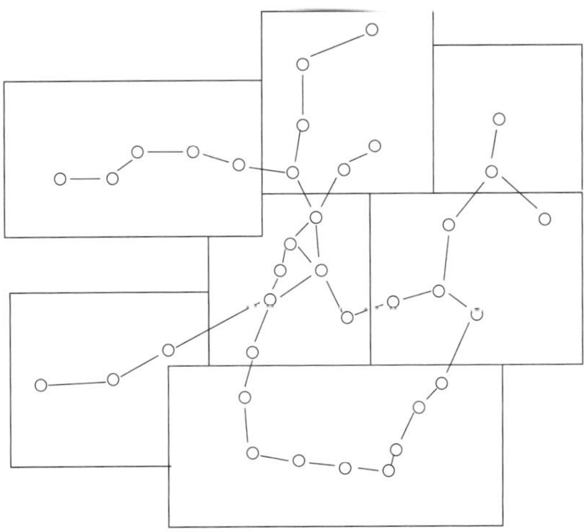

11

Introduction

Let us now deal with one more important issue in public transportation, namely the ticket prices for the customers. Designing a tariff system in public transportation is a complex real-world problem, that was brought to our attention by a regional public transportation company several years ago. In this part we present our studies and report on our practical experience in this area.

When using a bus or a train, a passenger usually has to pay for his trip. There are several possibilities for defining ticket prices in public transportation. The most popular ones are the distance tariff, the unit tariff and zone tariffs. In a *(counting) zone tariff system* the whole area of the public transportation company is divided into zones. To find out the ticket price for a trip, one counts the zones passed by the trip and reads off the price which only depends on the number of passed zones. Zone tariff systems are very popular at the moment, i.e., many transportation companies and traffic associations plan to introduce such a system.

When a public transportation company wants to change its tariff system to a zone tariff, it should be in such a way that the new system is still accepted by the customers and does not decrease the income of the company. A possible goal is to establish zones and zone prices such that the resulting ticket prices are as close as possible to the the current fares. This means that neither the public transportation company nor the customers will have major disadvantages when changing the current tariff system. Another goal can be to design a *fair* tariff system.

Given some preferred ticket prices (called *reference prices*), the *(counting) zone design problem* is to design a zone tariff system, i.e.,

- to design zones
- and zone prices for traveling within 1,2,3,... zones

such that the deviations between the resulting zone tariff and the given reference prices are as small as possible.

The reference prices are used to measure the quality of the new system. This is discussed in more detail in Section 11.4.

Chapter 11 is structured as follows: We first discuss the most common tariff systems. We then describe different zone design projects in which we applied our methodology in practice. Since there does not exist much literature on zone design problems, the literature review is rather short. Finally, we present a model for the zone design problem and three different possible objective functions.

11.1 Frequently Used Tariff Systems

We consider four different tariff systems: The distance tariff, the unit tariff, the zone tariff with arbitrary prices, and the counting zone tariff. Our goal will be to change a given tariff system to a counting zone tariff.

Distance Tariff

In a *distance tariff* system, the price for a trip depends on the length of the trip, given in kilometers. The longer the trip is, the higher is the fare. This system is usually considered as fair. To determine the ticket prices one needs the distance between each pair of stations, resulting in a matrix which in most cases is too large to be printed and put up at the stations. This makes a distance tariff inconvenient for the public transportation company and mysterious for the customers.
For the PTN with seven stations depicted in Figure 11.1, one needs the following 7×7 matrix to specify the distances (in kilometers) between each pair of stations.

$$D = \begin{pmatrix} 0 & 1 & 2 & 4 & 5 & 4 & 6 \\ 1 & 0 & 1 & 3 & 4 & 3 & 5 \\ 2 & 1 & 0 & 2 & 3 & 2 & 4 \\ 4 & 3 & 2 & 0 & 1 & 4 & 2 \\ 5 & 4 & 3 & 1 & 0 & 5 & 3 \\ 4 & 3 & 2 & 4 & 5 & 0 & 5 \\ 6 & 5 & 4 & 2 & 3 & 5 & 0 \end{pmatrix}$$

Furthermore, we need a table assigning a price for each trip length (for each ticket category), e.g.,

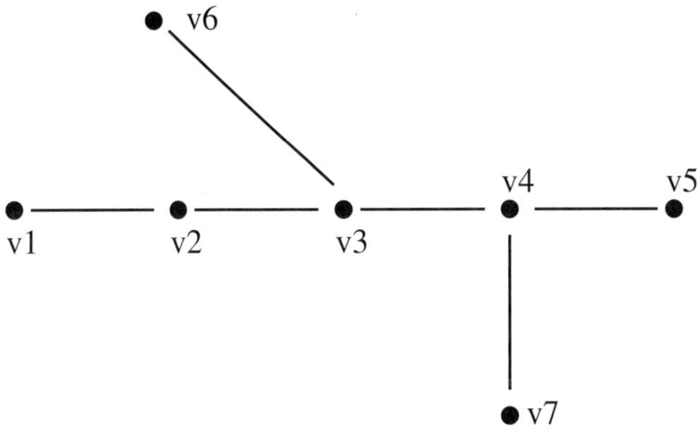

Fig. 11.1. A PTN to demonstrate the different tariff systems

kilometer	price
1	2
2	2
3	3
4	3
5	4
6	5
7	6

For example, to get the price for traveling from station v_1 to v_7 one has to look up $d_{v_1 v_7} = 6$ and can then read off a price of 5. Most railway transportation companies use a distance tariff system. Some years ago, distance tariffs were also used by most regional bus companies in Germany.

Unit Tariff

The simplest tariff system is the *unit tariff.* In this case all trips cost the same, independent of their length. A unit tariff is very easy to handle for the public transportation company, and it is easy to understand for the customers, since they only have to remember one price (in each ticket category). Taking, e.g., a price of 3 in the example of Figure 11.1 this means that the trip from v_1 to v_7 costs 3. On the other hand, also the short trip from v_1 to v_2 has a price of 3, which is annoying for the respective customers. In general, it often is not accepted that a short trip between two neighboring stations costs the same as a long trip through the whole system.

Unit tariff systems are used within metropolitan areas, or for small public transportation companies.

Zone Tariff

A model in between these two tariff systems is a *zone tariff system*. To establish a zone tariff, the whole area has to be divided into subregions (the *tariff zones*), see Figure 11.2 for the example of Figure 11.1. In this example, the zone partition is given by the following assignment.

station	zone
v_1	Z_1
v_2	Z_1
v_3	Z_2
v_4	Z_2
v_5	Z_3
v_6	Z_4
v_7	Z_3

The price for a trip in a zone tariff system depends only on the starting and the ending zone of the trip. We distinguish the following two different realizations of zone tariff systems.

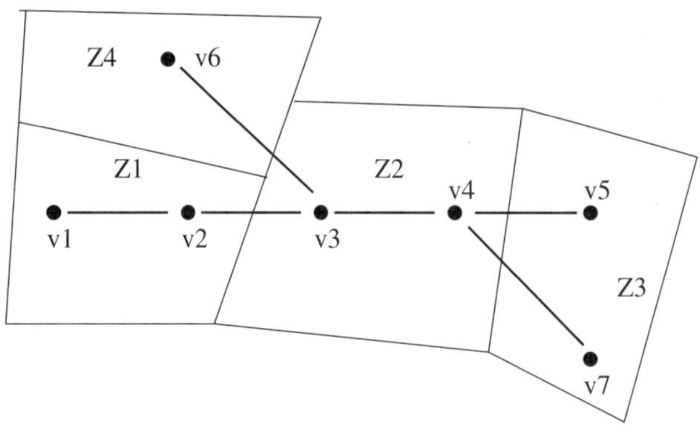

Fig. 11.2. The PTN with a zone partition of four zones

Zone tariff with arbitrary prices: If the price can be chosen arbitrarily for each pair of zones, we call the tariff system a *zone tariff with arbitrary prices*. The prices for each possible pair of zones are usually given in form of a matrix. Note that this matrix is much smaller than the matrix needed for describing a distance tariff. In the example of Figure 11.1 with the zone partition into four zones depicted in Figure 11.2, we only need the following 4×4 matrix to describe the tariff information instead of the 7×7 matrix for the distance tariff.

$$P = \begin{pmatrix} 1 & 2 & 4 & 3 \\ 2 & 1 & 3 & 2 \\ 4 & 3 & 3 & 5 \\ 3 & 2 & 5 & 1 \end{pmatrix}.$$

To find out the price for the trip from v_1 to v_7 one first has to find out that v_1 belongs to zone Z_1 and $v_7 \in Z_3$. Then the ticket price can be read off as entry $P_{Z_1 Z_3} = 4$. Note that the price for traveling within the same zone is not zero, and even need not be small, e.g., for traveling from v_5 to v_7.

This type of zone tariff is appropriate if any specialties should be modeled in the tariff system. An example of a zone tariff system with arbitrary prices can, for instance, be found north of San Francisco, USA, or within the city of Saarbrücken, Germany.

Counting zone tariff: A more popular variant of a zone tariff system is the *counting zone tariff system.* To know his fare in this system, a customer has to count how many zones his trip will pass and read off the price assigned to the number of passed zones. The prices in this system depend on the starting and the ending zone of the trip, but trips passing the same number of zones **must** have the same price. In the example depicted in Figure 11.2 this means that the price for traveling from Z_1 to Z_3 has to be the same as the prices from Z_1 to Z_4 and from Z_3 to Z_4. Counting the number of passed zones can be done easily by the customers themselves, if a map is available. For knowing the price for a trip they then only need the following (small) table.

number of zones	price
1	1
2	3
3	4

To find out the price for the trip from v_1 to v_7 we count that this trip passes through three zones and hence has a price of 4.

Because of their transparency, zone tariff systems are very popular. In Germany, nearly all tariff associations already have zone tariff systems or are currently introducing them, such that at the moment almost the whole local public transportation in Germany, including the commuter trains of *Deutsche Bahn*, applies counting zone tariff systems. Also in other countries, counting zone tariff systems are used, e.g., in the region south of San Francisco, USA.

Note that the set of direct connections E of the given public transportation network PTN is not necessary to design a zone tariff with arbitrary prices. For a counting zone tariff system, however, it is necessary to know the set of direct connections to count the number of passed zones for each possible trip.

We end this section by remarking that distance tariffs and unit tariffs can both be seen as special cases of a counting zone tariff system. While it is directly clear that a unit tariff is nothing else than a zone tariff system with only one zone, we need to add empty zones for modeling a distance tariff. If each kilometer along the direct connections is treated as a zone of its own, the number of passed zones of a trip equals the length of the trip in kilometers. The zone prices are hence nothing else than the prices assigned to the trip lengths in the distance tariff.

11.2 Application

Our first application was to design a zone tariff system with arbitrary prices for a regional bus company in the area of Kaiserslautern, see [Sch94a]. Since this time, other projects followed. We designed counting zone tariff systems, e.g., in the state Saarland and in a large part of the state Sachsen-Anhalt. The data we use for presenting numerical results within this part refers to our project in the Saarland. There are six public transportation companies operating in the Saarland. Before introducing the common zone tariff system, each of them had its own tariff system.

- Four public transportation companies already used a counting zone tariff system, but their zone prices for passing p zones and the structure of their zones were completely different, although they are partly operating in the same geographical region.
- *Deutsche Bahn* applied its distance tariff.
- There is also a public transportation company (serving the city of Saarland's capitol, Saarbrücken) which used a zone tariff with arbitrary prices.

The traffic association of the Saarland wanted to introduce one common counting zone tariff system which is now applied by all public transportation companies operating in the Saarland. The public transportation network in the Saarland consists of roughly 4000 stations, where a pre-clustering into 600 mini-zones is given. The goal was to design about 100 zones and install a counting zone tariff system in such a way that the difference between the current fares and the new ones is as small as possible. It also was important that the new income of each of the public transportation companies should not differ too much from the old income. While the old fare structure was known and therefore relatively easy to get, it is usually hard to get realistic data about customers' behavior. In our project in the Saarland this was solved by using the income data of each of the transportation companies and dividing the income with the help of available statistics among the origin-destination pairs used by the customers. However, since the data about customers are confidential, we use an unweighted variant of the zone design problem for presenting some of the numerical results.

11.3 Literature Review

In spite of the importance of the zone design problem there is hardly any literature on corresponding operations research models. The only papers we are aware of are motivated by a study of the author ([Sch94a]) about the zone design problem with arbitrary prices, see [HS95, Sch94b, Sch96, BK03]. These papers discussed complexity issues and heuristics for this type of problem, as well as exact solution procedures for special cases. For the counting zone design problem in which we count the number of zones there is, to the best of our knowledge, no literature dealing with suitable operations research models. Note that parts of this chapter have been published in [HS04], and that some first results in this area have been obtained in the diploma thesis [Pen97]. Related research includes simulation approaches for fare integration ([GM06]) and the determination of fares such as to maximize revenue ([BNP05]).

11.4 A Model for the Zone Design Problem

Let the public transportation network PTN $= (V, E)$ be a connected graph, where as usual, V refers to the set of stops and E represents the available direct rides without intermediate stops. Furthermore, let d_{uv} be given *reference prices* for traveling from station $u \in V$ to station $v \in V$. Our goal is to design zones and zone prices in such a way that they are a good approximation of the given reference prices.

- If d_{uv} is the current ticket price of the public transportation company, we aim to design a zone tariff system in which many customers will only have minor changes in their ticket prices, and hence will accept the new system. Also, the income of the company will not change much in this case.
- If d_{uv} represents a *fair* price like the distance tariff, the goal is to design a fair zone tariff system.
- We also allow any other possibility for d_{uv}.

If L denotes the number of planned zones, the *zone (planning) problem* identifies a partition

$$\mathcal{Z} = \{Z_1, Z_2, \ldots, Z_L\}$$

of V (i.e., $Z_i \subseteq V, i = 1, 2, \ldots, L$, the Z_i are pairwise disjoint and $\cup_{i=1}^{L} Z_i = V$). In the *fare (planning) problem* zone prices

$$c(p), p = 0, 1, 2, \ldots$$

are determined which depend only on the number of zones p in journey. Here $c(p)$ denotes the price for passing p zone borders. In particular, $c(0)$ gives the fare for traveling within the same zone (without passing any zone border), $c(1)$ is the price for passing one zone border, i.e., for going from one zone to an adjacent one, and so on. To evaluate some partition $\mathcal{Z}$ with a zone price vector c, we need to count the number of zones on a path from u to v. To this end, we need the following notation.

Notation 11.1. *For each pair of stations $u, v \in V$ let n_{uv} denote the minimal number of passed zone borders, when traveling from station u to station v.*

Two remarks are added.

- If there are several paths possible from u to v we take a path realizing the minimal value for n_{uv}. This means that we assume that customers choose the cheapest possibility for their journey. In regional transportation this is usually satisfied, since in most cases the shortest and the cheapest traveling possibility coincide. In long-distance rail transportation, however, this assumption need not be true, since other criteria like the traveling time and the number of changing activities become more important, and do sometimes lead to other paths and not to the cheapest one.
- Our definition of n_{uv} does not coincide with the usual notation of public transportation companies. When counting the number of zones n'_{uv} used within a trip from u to v in practice, the starting and the ending zone are both included, as we did in Section 11.1. This means that $n'_{uv} = n_{uv} + 1$. Referring once more to Figure 11.2 we see that for the trip from v_1 to v_7 the number $n_{v_1 v_7}$ of passed zone borders is 2, while $n'_{v_1 v_7} = 3$, i.e., three zones are touched.
  As it will turn out later, our model can be stated more simply by using the denotation n_{uv} instead of n'_{uv}.

The new ticket price for traveling from u to v is then given by

$$z_{uv} = \begin{cases} c(n_{uv}) & \text{if } u \neq v \\ 0 & \text{if } u = v. \end{cases}$$

Given the reference prices d_{uv} for a trip between stations u and v, the absolute deviation in ticket price is calculated by

$$|d_{uv} - z_{uv}| = |d_{uv} - c(n_{uv})|.$$

Recall that W_{uv} is the number of customers traveling from station u to station v. The minimization of the following three objective functions is of interest.

maximum absolute deviation: $b_{\max} = \max_{u,v \in V} W_{uv} |d_{uv} - z_{uv}|$
sum of absolute deviations: $b_1 = \sum_{u,v \in V} W_{uv} |d_{uv} - z_{uv}|$
sum of squared deviations: $b_2 = \sum_{u,v \in V} W_{uv} (d_{uv} - z_{uv})^2$

All three objectives are considered to be good models by practitioners. The first objective function, $b_{\max}$ with identical weights models the fact that the greatest deviation of ticket prices in the two different tariffs should be as small as possible. It gives a bound for changes in the ticket prices. In the weighted case, $b_{\max}$ minimizes the maximum deviation in the revenue of the company over all possible trips.

If W denotes the sum of all customers of the public transportation company, i.e., $W = \sum_{u,v \in V} W_{uv}$ then $\frac{b_1}{W}$ gives the average of all absolute deviations, and $\frac{b_2}{W}$ the average of all squared deviations in ticket prices. The objective function b_2 leads to a smaller percentage of strongly affected customers than b_1. Nevertheless, from our experience, b_1 is slightly better accepted by the practitioners than b_2. Furthermore, we point out that deviations in price increases and decreases are treated equally, such that the model reflects both the interests of the customers and of the transportation companies.

Before we discuss how to compute the values n_{uv} we introduce the following notation.

Notation 11.2. *Two zones $Z_k, Z_l \in \mathcal{Z}$ are called* **adjacent** *if there exist stops $u \in Z_k$, $v \in Z_l$ such that $\{u, v\} \in E$, i.e., with a direct ride in* PTN.

To obtain the numbers n_{uv} a shortest path algorithm, e.g., [Flo62, War62, Dij59] can be used according to one of the following models.

Station Graph Model: We use the public transportation network PTN $= (V, E)$, but introduce new weights c_{uv} for all $\{u, v\} \in E$, defined by

$$c_{uv} = \begin{cases} 0 & \text{if } u \text{ and } v \text{ are in the same zone} \\ 1 & \text{if } u \text{ and } v \text{ are in adjacent zones.} \end{cases}$$

The length of a shortest path between two stops equals the minimum number of passed zone borders. This approach will be needed later to update the zone distances in the greedy heuristic in Section 12.2.

Zone Graph Model: To reduce the size of the network we define the *zone graph* $G_Z = (\mathcal{Z}, E_Z)$ by

$$\begin{aligned} \mathcal{Z} &= \{Z_1, \ldots, Z_L\} \\ E_Z &= \{\{Z_k, Z_l\} : Z_k, Z_l \subset \mathcal{Z} \text{ and } Z_k \text{ and } Z_l \text{ are adjacent}\} \\ c_e &= 1 \quad \text{for all } e \in E_Z. \end{aligned}$$

For $u \in Z_k$ and $v \in Z_l$ we hence get the minimum number of passed zone borders n_{uv} on a trip from u to v as the length of a shortest path from Z_k to Z_l in G_Z.

The following example demonstrates the calculation of $b_{\max}$, b_1, and b_2. Let a PTN with a partition into three zones $Z_1 = \{v_1, v_2\}$, $Z_2 = \{v_3, v_4\}$, and $Z_3 = \{v_5\}$ be given (see Figure 11.3). Suppose that $W_{uv} = 1$ for all $u, v \in V, u \neq v$, i.e., $W = 20$. If we assume that the distance between any adjacent pair of nodes is 1, the matrix d_{uv} according to the distance tariff system may be

$$D = \begin{pmatrix} 0 & 1 & 2 & 3 & 4 \\ 1 & 0 & 1 & 2 & 3 \\ 2 & 1 & 0 & 1 & 2 \\ 3 & 2 & 1 & 0 & 1 \\ 4 & 3 & 2 & 1 & 0 \end{pmatrix}.$$

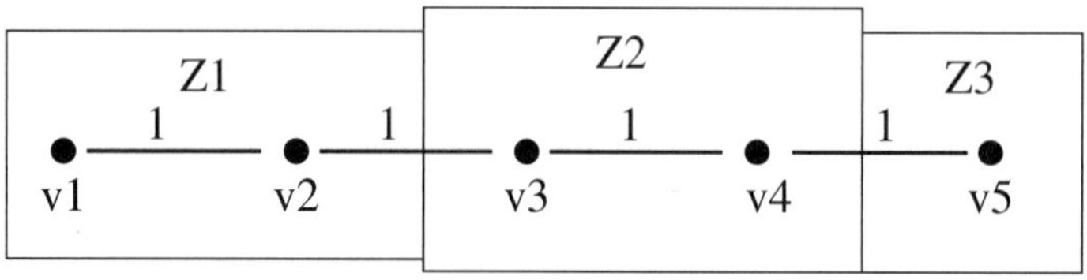

Fig. 11.3. The PTN of the example

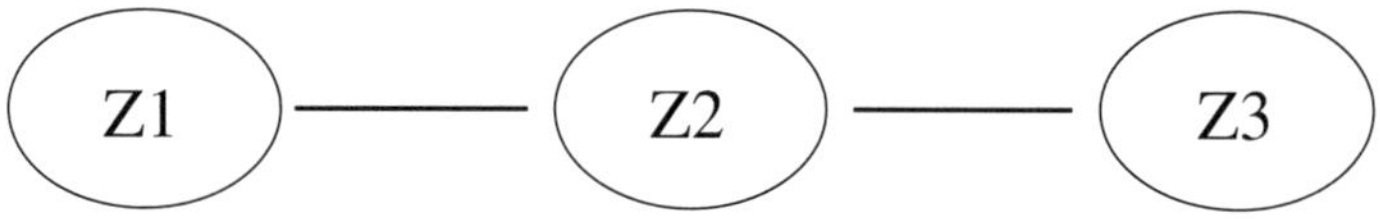

Fig. 11.4. The corresponding zone graph G_Z of the example

The corresponding zone graph G_Z consists of three nodes (see Figure 11.4). The number of passed zone borders between stations u and v is then given by

$$N = \begin{pmatrix} 0 & 0 & 1 & 1 & 2 \\ 0 & 0 & 1 & 1 & 2 \\ 1 & 1 & 0 & 0 & 1 \\ 1 & 1 & 0 & 0 & 1 \\ 2 & 2 & 1 & 1 & 0 \end{pmatrix}.$$

Suppose the new fares for passing $p = 0, 1$, or 2 zone borders are given by

$$\begin{aligned} c(0) &= 0.5 \\ c(1) &= 1 \\ c(2) &= 1.5. \end{aligned}$$

Then the new ticket prices can be calculated as

$$Z = \begin{pmatrix} 0 & 0.5 & 1 & 1 & 1.5 \\ 0.5 & 0 & 1 & 1 & 1.5 \\ 1 & 1 & 0 & 0.5 & 1 \\ 1 & 1 & 0.5 & 0 & 1 \\ 1.5 & 1.5 & 1 & 1 & 0 \end{pmatrix}.$$

The deviations between the reference prices d_{uv} and the new ticket prices z_{uv} are

$$D - Z = \begin{pmatrix} 0 & 0.5 & 1 & 2 & 2.5 \\ 0.5 & 0 & 0 & 1 & 1.5 \\ 1 & 0 & 0 & 0.5 & 1 \\ 2 & 1 & 0.5 & 0 & 0 \\ 2.5 & 1.5 & 1 & 0 & 0 \end{pmatrix}.$$

and finally the objective function values can be computed as

$$\begin{aligned} b_{\max} &= 2.5 \\ b_1 &= 20 \\ b_2 &= 32, \end{aligned}$$

i.e., the maximum absolute deviation is 2.5, the average absolute deviation is 1, and the average squared deviation is 1.6.

12

Finding Zones and Zone Prices

In this chapter we deal with the counting zone design problem. We assume that the public transportation network and the necessary data about the customers in form of an OD-matrix is given. With the notation introduced in the previous chapter, the *counting zone design problem* is given as follows.

(ZD-b)

Given the PTN, *reference prices* d_{uv}, *an OD-matrix with entries* W_{uv}, *and* $L \in \mathbb{N}$, *find a partition of* V *into* L *zones* $Z_k, k = 1, \ldots, L$ *and zone prices* $c(p), p = 1, \ldots, L$ *such that an objective function*

$$b \in \{b_{\max}, b_1, b_2\}$$

is minimized, where $b_{\max}, b_1,$ *and* b_2 *are given as follows,*

$$\begin{aligned} b_{\max}(\mathcal{Z}, c) &= \max_{u,v \in V} W_{uv} |d_{uv} - c(n_{uv})| \\ b_1(\mathcal{Z}, c) &= \sum_{u,v \in V} W_{uv} |d_{uv} - c(n_{uv})| \\ b_2(\mathcal{Z}, c) &= \sum_{u,v \in V} W_{uv} (d_{uv} - c(n_{uv}))^2. \end{aligned}$$

Chapter 12 is structured as follows: In the first section of this chapter we deal with the *fare problem*, i.e., we assume that the zone partition $\mathcal{Z}$ is already given and show how to find optimal zone prices c. We present closed form solutions for each of the three objective functions. Then we turn our attention to the *zone problem* in which we want to find the zones and exemplarily investigate the case of (ZD-$b_{\max}$). We show that this problem is NP-hard and we therefore propose three heuristic algorithms. We also discuss useful extensions to apply the model in practice.

12.1 The Fare Problem

In this section we solve the fare problem with respect to a given zone partition. This means, we assume that the partition $\mathcal{Z}$ is already fixed and deal with determining the zone prices

$$c(p), p = 0, 1, 2, \ldots$$

Our first result shows that a closed form solution is possible for each of the three objectives $b_{\max}, b_1$, and b_2. To state the result we first introduce the following denotation, restricting the fare problem to the price for passing p zone borders for any fixed p.

Notation 12.1. *Given $p \in \{0, 1, \ldots, L\}$ let*

$$M_p = \{(u, v) : u, v \in V, u \neq v, \text{ and } n_{uv} = p\}.$$

Moreover, $W^p = \sum_{(u,v)\in M_p} W_{uv}$ denotes the sum of all weights belonging to pairs of stations in the set M_p.

Note that $u \neq v$ is only necessary in the definition of M_0.

Theorem 12.2. *Let $\mathcal{Z} = \{Z_1, Z_2, \ldots, Z_L\}$ be a given zone partition and let d_{uv} be given reference prices. In order to minimize $b_{\max}$, b_1, and b_2 we choose for all $p = 0, 1, \ldots, L$,*

a)

$$c^*_{\max}(p) := \max_{(u,v)\in M_p} d_{uv} - \frac{z^*_p}{W_{uv}}$$

where z^*_p is defined as

$$z^*_p = \max_{\substack{u_1,v_1,u_2,v_2:\\(u_1,v_1),(u_2,v_2)\in M_p}} \frac{W_{u_1v_1} W_{u_2v_2}}{W_{u_1v_1} + W_{u_2v_2}} (d_{u_1v_1} - d_{u_2v_2})$$

b)

$$c^*_1(p) := \text{median}\{\underbrace{d_{uv}, \ldots, d_{uv}}_{W_{uv} \text{ times}} : (u, v) \in M_p\}$$

c)

$$c^*_2(p) := \frac{1}{W^p} \sum_{(u,v)\in M_p} W_{uv} d_{uv}.$$

Proof. Given the zone partition $\mathcal{Z}$ we have to find fares $c(p) \in \mathbb{R}$ for all $p = 0, 1, \ldots,$ minimizing $b_{\max}$, b_1, and b_2, respectively. First we note that each of the three objective functions can be separated into at most $L + 1$ independent subproblems, $K_{\max}(p)$, $K_1(p)$, and $K_2(p)$, respectively (for $p = 0, 1, \ldots, L$).

$$\begin{aligned}
b_{\max} &= \max_{u,v \in V} W_{uv}|d_{uv} - z_{uv}| \\
&= \max_{p=0,1,\dots,L} \max_{m \in M_p} W_m|d_m - c(p)| =: \max_{p=0,1,\dots,L} K_{\max}(p) \\
b_1 &= \sum_{u,v \in V} W_{uv}|d_{uv} - z_{uv}| \\
&= \sum_{p=0}^{L} \sum_{m \in M_p} W_m|d_m - c(p)| =: \sum_{p=0}^{L} K_1(p) \\
b_2 &= \sum_{u,v \in V} W_{uv}(d_{uv} - z_{uv})^2 \\
&= \sum_{p=0}^{L} \sum_{m \in M_p} W_m(d_m - c(p))^2 =: \sum_{p=0}^{L} K_2(p).
\end{aligned}$$

Consequently, to minimize $b_{\max}$, b_1, and b_2 we determine the optimal fare $c(p)$ for $p = 0, 1, \dots, L$ separately, in each of the three objective functions.

For $b_{\max}$: For all $p = 0, 1, \dots, L$ the problem of finding a value $c(p)$ that minimizes

$$K_{\max}(p) = \max_{m \in M_p} W_m|d_m - c(p)|$$

is well-known from location theory when locating a point on a line such that the maximum distance to a given set of existing facilities on the same line is minimized. The proof for the formula given in part a) of the theorem can therefore be found in the location literature, see e.g. [LMW88, Ham95]. Note that it is also known that for the optimal prices $c^*_{\max}$ we have

$$K_{\max}(p) = z^*_p. \tag{12.1}$$

For b_1: Since

$$K_1(p) = \min \sum_{m \in M_p} W_m|d_m - c(p)|$$

is a one-dimensional, piecewise linear and convex function, its minimization is known in statistics (see e.g., [Hay81]) and in location theory as the one-dimensional median problem (see, e.g. [Ham95, Pla95]). It is shown that the above problem is solved by the so-called *weighted median* of the set $\{d_m : m \in M_p\}$, i.e., by any real number $c = c^*_1(p)$ which satisfies

$$\sum_{m:d_m<c} W_m \le \frac{W^p}{2} \quad \text{and}$$

$$\sum_{m:d_m>c} W_m \le \frac{W^p}{2}.$$

For b_2: Here we have to minimize $K_2(p)$, i.e.,

$$\min \sum_{m \in M_p} W_m (d_m - c(p))^2.$$

Using the theorem of Steiner (see e.g. [SV74]) of statistics, we note that the weighted mean of the values in $\{d_m : m \in M_p\}$ is the unique optimal solution for $c(p)$. □

To demonstrate the results of Theorem 12.2 we continue the example of Section 11.4, depicted in Figure 11.3 (see page 215). First we determine the relations for the sets M_p.

$$\begin{aligned} M_0 &= \{(v_1, v_2), (v_2, v_1), (v_3, v_4), (v_4, v_3)\} \\ M_1 &= \{(v_1, v_3), (v_3, v_1), (v_1, v_4), (v_4, v_1), (v_2, v_3), (v_3, v_2), \\ &\quad (v_2, v_4), (v_4, v_2), (v_3, v_5), (v_5, v_3), (v_4, v_5), (v_5, v_4)\} \\ M_2 &= \{(v_1, v_5), (v_5, v_1), (v_2, v_5), (v_5, v_2)\}. \end{aligned}$$

The optimal values for the zone prices with respect to the objective functions $b_{\max}$, b_1, and b_2 can hence be calculated as

zones	$c^*_{\max}$	c^*_1	c^*_2	example
0	1	1	1	0.5
1	2	2	$\frac{11}{6}$	1
2	3.5	3	3.5	1.5

where the last column contains the prices from the example. Finally, the resulting objective values are:

objective function	$c^*_{\max}$	c^*_1	c^*_2	example
$b_{\max}$	**1**	1	1.167	2.5
b_1	8	**8**	8.667	20
b_2	7	8	**6.722**	32

Note that the best objective value for $b_{\max}$ is attained for the prices $c^*_{\max}$, the best objective value for b_1 for the prices c^*_1, and the minimum of b_2 is attained at b^*_2.

Calculating the objective function by using the optimal fares according to Theorem 12.2 implies the following results.

Corollary 12.3. *Given a zone partition* $\mathcal{Z} = \{Z_1, Z_2, \ldots, Z_L\}$ *and reference prices* d_{uv}*, the optimal values of the objective functions are given as follows.*

a)
$$b^*_{\max} = \max_p z^*_p$$

b)
$$b^*_1 = \sum_p \left(\sum_{\substack{(u,v)\in M_p: \\ d_{uv} > c^*_1(p)}} W_{uv}(d_{uv} - c^*_1(p)) + \sum_{\substack{(u,v)\in M_p: \\ d_{uv} < c^*_1(p)}} W_{uv}(c^*_1(p) - d_{uv}) \right)$$

c)
$$b^*_2 = \sum_p \operatorname{Var}\{d_{uv} : (u,v) \in M_p\},$$

where Var *denotes the variance of the set.*

In practice, restrictions on the new fares are often given; sometimes there even exist "politically" desired fares for the number of zones in a journey that have to be realized. With the help of Corollary 12.3 one can easily calculate the increase of the objective functions when using such given fares instead of the optimal ones.

Another important consequence of Corollary 12.3 is that for the objective function $b_{\max}$ the optimal fares $c^*_{\max}(p)$ are not needed for calculating the optimal objective value for a given zone partition. This will be needed in the next section when we are going to optimize the zone partition with respect to $b_{\max}$. If, additionally, $b_{\max}$ is used in the unweighted case, i.e. with $W_{uv} = 1$ for all $u, v \in V$, we can further simplify Theorem 12.2 and Corollary 12.3. Recall that

$$c^*_{\max}(p) = \max_{(u,v)\in M_p} d_{uv} - \frac{z^*_p}{W_{uv}}$$
$$K_{\max}(p) = \max_{(u,v)\in M_p} W_{uv}|d_{uv} - c(p)|$$

and for the optimal zone prices $c^*_{\max}$ we have

$$K_{\max}(p) = z^*_p, \text{ see (12.1).}$$

Corollary 12.4. *Let a zone partition* $\mathcal{Z} = \{Z_1, Z_2, \ldots, Z_L\}$ *and reference prices* d_{uv} *be given and assume equal weights* $W_{uv} = 1$ *for all* $(u,v) \in V \times V$*. Then the optimal fares* $c^*_{\max}(p)$ *and the corresponding objective values* $b^*_{\max} = b_{\max}(\mathcal{Z}, c^*_{\max})$*, and* $K^*_{\max}$ *are given by*

$$c^*_{\max}(p) = \frac{1}{2}\left(\max_{(u,v)\in M_p} d_{uv} + \min_{(u,v)\in M_p} d_{uv} \right)$$

$$K^*_{\max}(p) = \frac{1}{2}\left(\max_{(u,v)\in M_p} d_{uv} - \min_{(u,v)\in M_p} d_{uv}\right)$$

$$b^*_{\max} = \frac{1}{2}\max_{p=1,\ldots,L}\left(\max_{(u,v)\in M_p} d_{uv} - \min_{(u,v)\in M_p} d_{uv}\right).$$

Proof. We calculate z^*_p as

$$\begin{aligned} z^*_p &= \max_{m_1,m_2\in M_p} \frac{W_{m_1}W_{m_2}}{W_{m_1}+W_{m_2}}(d_{m_1}-d_{m_2}) \\ &= \frac{1}{2}\left(\max_{m\in M_p} d_m - \min_{m\in M_p} d_m\right) \end{aligned}$$

and consequently,

$$\begin{aligned} c^*_{\max}(p) &= \max_{m\in M_p} d_m - \frac{z^*_p}{W_m} \\ &= \max_{m\in M_p}\left(d_m - \frac{1}{2}\max_{\tilde{m}\in M_p} d_{\tilde{m}} + \frac{1}{2}\min_{\tilde{m}\in M_p} d_{\tilde{m}}\right) \\ &= \frac{1}{2}\left(\max_{m\in M_p} d_m + \min_{m\in M_p} d_m\right). \end{aligned}$$

From (12.1) we know that $z^*_p = K^*_{\max}(p)$; hence the remaining parts follow immediately from part a) of Corollary 12.3. □

We remark that for the zone design problem with arbitrary prices, similar results can be derived (see [HS95, Sch94a]).

12.2 The Maximum Deviation Zone Design Problem

The consequence of the results of Section 12.1 is that we can concentrate on finding the zones, since the zone pricing follows easily from the choice of the objective function. We therefore turn our attention to the zone (planning) problem. In particular, we now focus on the maximum deviation problem (ZP-$b_{\max}$) in the unweighted case.

Notation 12.5. *For a given zone partition $\mathcal{Z} = \{Z_1, Z_2, \ldots, Z_L\}$ and reference prices d_{uv} let $b^*_{\max}$ and $K^*_{\max}$ denote the optimal objective function values of the fare problem according to Corollary 12.4.*

A first observation deals with the monotonicity of the objective function dependent on the number of planned zones L. While it is easy to see that for the zone design problem with arbitrary prices (see Section 11.1) all three objectives are monotone in L, this is not true for the zone design problem with counting zones, as Figure 12.1 shows. The PTN in this example consists of eight nodes. Let us assume that $W_{uv} = 1$ for all pairs of nodes u, v. The reference prices are given as weights between any two adjacent nodes, as shown in the figure. Between any other pair of nodes the reference prices are given as the sum of the weights along a shortest path connecting the nodes. For the (unweighted) max absolute deviation problem, we first calculate the objective value $b^*_{\max}$ for the graphed solution $\mathcal{Z}$ by Corollary 12.4.

$$K^*_{\max}(0) = \frac{1}{2}(1-1) = 0$$
$$K^*_{\max}(1) = \frac{1}{2}(102-100) = 1$$
$$K^*_{\max}(2) = \frac{1}{2}(202-201) = \frac{1}{2},$$

hence $b^*_{\max} = 1$. In any solution with $L = 5$ nonempty zones we will at least have one zone containing only one single station. Then

$$\begin{aligned} K^*_{\max}(1) &\geq \frac{1}{2}(100-1) \\ &= \frac{99}{2} > 1, \end{aligned}$$

leading to a strictly higher objective value than the graphed solution for $L = 4$. Similarly, we can verify that also allowing empty zones will not yield a better objective value than 1.

Unfortunately, the zone design problem with $b_{\max}$ is NP-hard, even in the unweighted case.

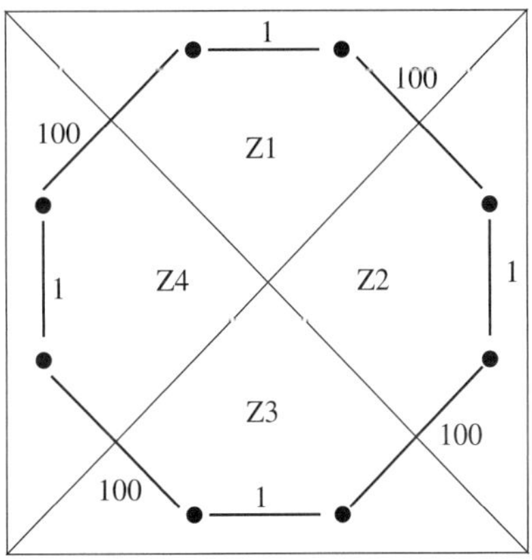

Fig. 12.1. An example in which the optimal solution for four zones is smaller than the optimal solution for five zones.

Theorem 12.6. *The zone design problem with counting zones and objective function $b_{\max}$ is NP-hard for all fixed $L \geq 3$.*

Proof. We use a reduction to the problem *partition into cliques* for $L = 3$ cliques, which is NP-hard (see problem [GT15] in [GJ79b]). It is given as follows.

(Partition into cliques) Given a graph $\mathcal{G} = (\mathcal{V}, \mathcal{E})$, does there exists a partition of $\mathcal{V}$ into three node sets $\mathcal{V}_1, \mathcal{V}_2$, and $\mathcal{V}_3$, such that the induced subgraphs $\mathcal{G}_1, \mathcal{G}_2$, and $\mathcal{G}_3$ are complete?

To reduce this problem to an unweighted zone design problem, we define the public transportation network as a complete graph PTN $= (V, E)$ by

$$V = \mathcal{V} \cup \{a_1, a_2, a_3, b_1, b_2, b_3\} \text{ and}$$
$$E = \{\{k, l\} : k, l \in V\}, k \neq l.$$

Furthermore, let $W_{uv} = 1$ for all $u, v \in V$ and define the reference prices as follows.

$$d_{uv} = \begin{cases} 1 & \text{if } \{u,v\} \in \mathcal{E} \\ 1 & \text{if } u \in \mathcal{V}, v \notin \mathcal{V} \\ \frac{1}{2} & \text{if there exists } i = 1,2,3 \text{ such that } \{u,v\} = \{a_i, b_i\} \\ 2 & \text{if } u, v \in \mathcal{V}, \{u,v\} \notin \mathcal{E} \\ 2 & \text{if } u, v \notin \mathcal{V}, \{u,v\} \neq \{a_i, b_i\} \text{ for all } i = 1,2,3. \end{cases}$$

Claim: $\mathcal{G}$ can be partitioned into three cliques if and only if the zone design problem in PTN has a solution $\mathcal{Z}$ with three zones and with $b^*_{\max} < \frac{3}{4}$.

$\Longrightarrow$: Let $\mathcal{V} = \mathcal{V}_1 \cup \mathcal{V}_2 \cup \mathcal{V}_3$ be the partition of $\mathcal{G}$ into cliques. Define $Z_i = \mathcal{V}_i \cup \{a_i, b_i\}$ for $i = 1, 2, 3$. Using Corollary 12.4 we calculate for the zone partition $\mathcal{Z} = \{Z_1, Z_2, Z_3\}$ that

$$K^*_{\max}(0) = \frac{1}{2}\left(1 - \frac{1}{2}\right) = \frac{1}{4}$$
$$K^*_{\max}(1) = \frac{1}{2}(2 - 1) = \frac{1}{2},$$

such that we get $b^*_{\max} < \frac{3}{4}$.

$\Longleftarrow$: Let Z_1, Z_2, Z_3 be a partition of V with $b^*_{\max} < \frac{3}{4}$. Define $\mathcal{V}_i = Z_i \cap V$ for $i = 1, 2, 3$. First we prove that we can rename the Z_i such that $a_i, b_i \in Z_i$ for $i = 1, 2, 3$. This can be done by using Corollary 12.4 to verify the following.

1. If a_i, b_i and another a_j (or b_j), $j \neq i$ belong to a single common zone, then

$$\begin{aligned} b^*_{\max} &\geq K^*_{\max}(0) \\ &\geq \frac{1}{2}\left(d_{a_i a_j} - d_{a_i b_i}\right) \\ &= \frac{1}{2}\left(2 - \frac{1}{2}\right) = \frac{3}{4}. \end{aligned}$$

2. If a_i and b_i do not belong to the same zone, then they belong to two different zones Z_x, Z_y, such that
   - either a_j (or b_j) is contained in Z_x (or Z_y), hence

$$\begin{aligned} b^*_{\max} &\geq K^*_{\max}(1) \\ &\geq \frac{1}{2}\left(d_{a_j b_i} - d_{a_i b_i}\right) \\ &= \frac{1}{2}\left(2 - \frac{1}{2}\right) = \frac{3}{4}. \end{aligned}$$

   - or no other a_j, b_j is contained in Z_x and Z_y, implying that the remaining zone satisfies condition 1; hence again $b^*_{\max} \geq \frac{3}{4}$.

Now let $u, v \in \mathcal{V}_i$. We have to show that the edge $\{u, v\} \in \mathcal{E}$. Assume the contrary, i.e., $d_{uv} = 2$. But this yields

$$\begin{aligned} b^*_{\max} &\geq K^*_{\max}(0) \\ &\geq \frac{1}{2}\left(d_{uv} - d_{a_i b_i}\right) \\ &\geq \frac{1}{2}\left(2 - \frac{1}{2}\right) = \frac{3}{4}, \end{aligned}$$

(again using Corollary 12.4), a contradiction.

For more than three zones, the proof can be done analogously with a reduction to partition into $L > 3$ cliques. □

We now discuss three heuristics for finding good zone partitions. As a motivation, we first present the following two observations for getting upper and lower bounds on the objective value $b_{\max}$ of (TD-$b_{\max}$).

Lemma 12.7. *For any zone partition $\mathcal{Z}$ we have*

$$b^*_{\max} \leq \frac{1}{2}\left(\max_{u,v \in V} d_{uv} - \min_{u,v \in V} d_{uv}\right).$$

Proof. For any zone partition $\mathcal{Z}$ and any integer p we have that

$$\begin{aligned} K^*_{\max}(p) &= \frac{1}{2}\left(\max_{(u,v) \in M_p} d_{uv} - \min_{(u,v) \in M_p} d_{uv}\right) \\ &\leq \frac{1}{2}\left(\max_{u,v \in V} d_{uv} - \min_{u,v \in V} d_{uv}\right). \end{aligned}$$

Hence $b^*_{\max} = \max_{p=1,\ldots,L} K^*_{\max}(p)$ also satisfies this inequality. □

For the next bound we define the following two sets.

Notation 12.8. *For a given zone partition $\mathcal{Z}$, let*

$$\begin{aligned} \text{INT} &= \{(u,v) : \{u,v\} \in E \textit{ and there exists } Z \in \mathcal{Z} \textit{ with } u,v \in Z\}, \\ \text{BET} &= \{(u,v) : \{u,v\} \in E \textit{ and there exist } Z_1, Z_2 \in \mathcal{Z} \textit{ with} \\ &\qquad Z_1 \neq Z_2 \textit{ and } u \in Z_1, v \in Z_2\}, \end{aligned}$$

i.e., INT is the set of relations belonging to edges with endpoints in the interior of a zone, and BET contains relations with endpoints in adjacent zones (between zones).

Lemma 12.9. *Let a zone partition $\mathcal{Z}$ be given, together with the sets* INT *and* BET. *Then the following statements hold.*

1. $b^*_{\max} \geq \frac{1}{2}\left(\max_{(u,v)\in \text{INT}} d_{uv} - \min_{(u,v)\in \text{INT}} d_{uv}\right).$
2. $b^*_{\max} \geq \frac{1}{2}\left(\max_{(u,v)\in \text{BET}} d_{uv} - \min_{(u,v)\in \text{BET}} d_{uv}\right).$

Proof.

1. Since INT $\subseteq M_0$ we have that

$$\begin{aligned} \min_{(u,v)\in M_0} d_{uv} &\leq \min_{(u,v)\in \text{INT}} d_{uv}, \text{ and} \\ \max_{(u,v)\in M_0} d_{uv} &\geq \max_{(u,v)\in \text{INT}} d_{uv}. \end{aligned}$$

Hence we obtain

$$\begin{aligned} b^*_{\max} &\geq K^*_{\max}(0) \\ &= \frac{1}{2}\left(\max_{(u,v)\in M_0} d_{uv} - \min_{(u,v)\in M_0} d_{uv}\right) \\ &\geq \frac{1}{2}\left(\max_{(u,v)\in \text{INT}} d_{uv} - \min_{(u,v)\in \text{INT}} d_{uv}\right). \end{aligned}$$

2. Analogously, since BET $\subseteq M_1$ we have that

$$\begin{aligned} \min_{(u,v)\in M_1} d_{uv} &\leq \min_{(u,v)\in \text{BET}} d_{uv}, \text{ and} \\ \max_{(u,v)\in M_1} d_{uv} &\geq \max_{(u,v)\in \text{BET}} d_{uv} \end{aligned}$$

and the required result is then obtained as before by using $b^*_{\max} \geq K^*_{\max}(1)$. □

Lemma 12.9 suggests a zone design in which edges with high weights are collected in BET and edges with small weights in INT or vice versa. To be more specific, let Diam be the maximal diameter over all zones in the zone partition $\mathcal{Z}$. Assuming that edge weights along a path are additive, we get

$$\begin{aligned} K^*_{max}(p) &= \frac{1}{2}\left(\max_{(u,v)\in M_p} d_{uv} - \min_{(u,v)\in M_p} d_{uv}\right) \\ &\leq \frac{1}{2}\left((p+1)\mathrm{Diam} + p\left(\max_{(u,v)\in \mathrm{BET}} d_{uv} - \min_{(u,v)\in \mathrm{BET}} d_{uv}\right)\right) \end{aligned}$$

yielding that the maximal diameter Diam should be small, and consequently edges with large weights should be in BET while edges with small weights should be in INT. Following these considerations, we present three heuristics for the zone design problem with counting zones.

Algorithms Based on Clustering Theory

The first algorithm is based on ideas from clustering theory and here in particular on the SAHN (sequential agglomerative hierarchical non-overlapping) algorithms, see, e.g., [DO74]. The idea is to start with $n = |V|$ zones, each of them containing one single station and to combine in each step the two closest zones to a new one. Depending on the particular definition of the distance between two zones, different algorithms can be obtained. Two of them have been applied to the zone design problem: *single linkage* and *complete linkage*.

Algorithm 23: Zone design using SAHN-algorithms

Input: PTN, reference prices d_{uv}, $L \in \mathbb{N}$.
Output: Zone partition with L nonempty zones.
Step 1. Start with a partition $\mathcal{Z}$ consisting of $|V|$ zones each of them containing a single station.
Let $d(Z_u, Z_v) = d_{uv}$ for all zones $Z_u, Z_v \in \mathcal{Z}$.
Step 2. Determine two zones $Z_u \neq Z_v \in \mathcal{Z}$ with minimum distance $d(Z_u, Z_v)$.
Step 3. Join Z_u and Z_v to a new zone Z_k and get a new partition $\mathcal{Z}$.
Step 4. Calculate the new distances to all $Z \in \mathcal{Z}$:
$d(Z_k, Z) = \frac{1}{2}\left(d(Z_u, Z) + d(Z_v, Z) + \mathsf{c}|d(Z_u, Z) - d(Z_v, Z)|\right)$
Step 5. If the number of planned zones is attained, then Stop,
Output: $\mathcal{Z}$,
else goto 2.

The parameter c in step 4 determines the formula for calculating the distance between two zones. In the context of the zone design problem, we have used

- $c = -1$ for the single linkage algorithm and
- $c = 1$ for the complete linkage algorithm.

The interpretation is the following: In the single linkage algorithm, the distance between two zones is defined as the smallest distance between elements of them, and consequently we join along a shortest edge in each step. In the complete linkage algorithm, the distance between two zones is defined as the maximum distance between their elements. Hence, in each step complete linkage tries to minimize the maximum diameter of the zones.

Greedy Approach

This approach is a variant of the SAHN algorithms discussed above, but with more emphasis on the specific structure of the zone design problem. Using the basics of Algorithm 23, we calculate for all edges $\{Z_u, Z_v\}$ the objective value $b^{uv}_{\max}$ when contracting $\{Z_u, Z_v\}$ of the current zone graph. Finally, we contract the edge with smallest increase in the objective function. This is rather time consuming, but as we will show in the next section, leads to very good results in practice. The formulation of the greedy approach is the following:

Algorithm 24: Zone design by greedy approach

Input: PTN, reference prices d_{uv}, $L \in \mathbb{N}$.
Output: Zone partition with L zones.
Step 1. Start with a partition $\mathcal{Z}$ consisting of $|V|$ zones each of them containing a single station.
Step 2. For each edge $\{Z_u, Z_v\}$ in E_Z contract Z_u and Z_v temporally and calculate $b^{uv}_{\max}$ for the resulting zone partition.
Step 3. Contract the edge $\{Z_{u^0}, Z_{v^0}\}$ permanently, where

$$b^{u^0,v^0}_{\max} = \min_{\{Z_u,Z_v\}\in E_Z} b^{uv}_{\max},$$

and get a new partition $\mathcal{Z}$.
Step 3. If the graph has L nodes, then Stop,
Output: $\mathcal{Z}$,
else goto 2.

Spanning Tree Approach

The idea of the following heuristic is to determine a set of edges BET which contains mainly edges with high weights. To this end, we start with a large zone containing all stations and separate it into smaller zones. This is done by deleting edges in a spanning tree until the required number of zones is attained. The heuristic is formulated next.

Algorithm 25: Zone design by spanning tree approach

Input: PTN, reference prices d_{uv}, $L \in \mathbb{N}$.
Output: Zone partition with L non-empty zones.
Step 1. Find a maximum spanning tree T in the complete graph with edge weights d_{uv}.
Step 2. Delete the $L-1$ largest edges of T and get a forest with L components.
Step 3. Output: Zones are the connected components.

Note that in trees, the spanning tree approach is equivalent to the single linkage algorithm of clustering theory. In general graphs, it is always possible to find a spanning tree such that omitting its $L-1$ largest edges leads to the same result as single linkage. However, if we start with a spanning tree with maximal weight (which performed best in practice) the spanning tree approach differs significantly from single linkage.

Comparison of the Heuristics

We tested our algorithms on the data described above. The results of Algorithms 23, 24, and 25 are shown in Figure 12.2. This figure shows the objective values $b_{\max}$ produced by the heuristics for any number of possible zones from 1 to 600. The objective values refer to a single trip ticket for an adult, given in €. It turns out that in this practical application the greedy heuristic (Algorithm 24) is the clear winner in terms of the objective value: it generated the best results for any number of desired zones. On the other hand, the running time of Algorithm 24 for all possible numbers of zones, i.e., from $L = 1, \ldots, 600$ was nearly two weeks altogether in our first implementation (on an AixJ90). The spanning tree approach (Algorithm 25) and the single linkage algorithm (Algorithm 23) both needed only a few hours, but the results are much less convincing regarding the objective value $b_{\max}$ again. For a small number L of desired zones, single linkage did better than the spanning tree approach, while for a higher number of planned zones it was the other way round. This is due to the fact that the spanning tree approach starts with only one zone, while single linkage starts with 600 zones.

On a subset consisting of only 400 stations (or 54 mini-zones) the heuristics have also been tested. In this smaller setting the running times of Algorithms 23 and 25 were within seconds, and also Algorithm 24 needed only two minutes to obtain again the clearly best results. The results for nine zones

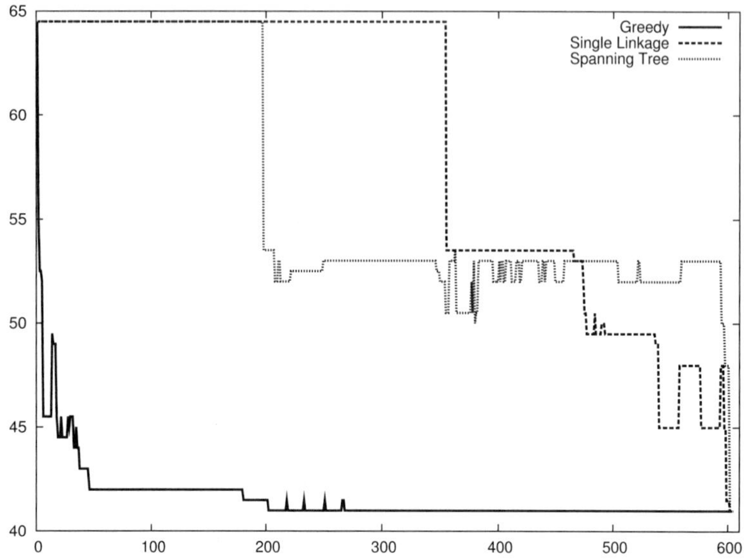

Fig. 12.2. Comparison of the heuristics: $b_{\max}$ graphed for any number L of planned zones.

are shown graphically in Figures 12.3,12.4, and 12.5. Figure 12.3 shows a suggestion for a zone partition which is due to the political districts in this area of the Saarland. The objective value for this zone partition is $b_{\max} = 2.63$ €, i.e., there exists a customer that will have a difference of 2.63 € between his current fare and the new one. The result of the single linkage algorithm for nine zones is shown in Figure 12.4. As it is reported also in the literature (see, e.g., [DO74]), single linkage tends to form one large zone and a lot of smaller zones surrounding it. This behavior is also shown in Figure 12.4. The objective value of the graphed zone partition is 2.56 €. The objective value in the spanning tree approach also was 2.56 € for nine zones, but without these big differences in the sizes of the zones. The best results, however, were obtained by the greedy approach with an objective value of only 1.92 €. The corresponding zone partition is shown in Figure 12.5.

12.3 Extensions for Real-world Problems

For evaluating tariff zones in more detail, we use our own software WabPlan [SS99]. For any given zone partition and zone prices it provides not only the objective values of $b_{\max}$, b_1, and b_2, but also the expected income for each of the transportation companies in each ticket category. This is done by evaluating

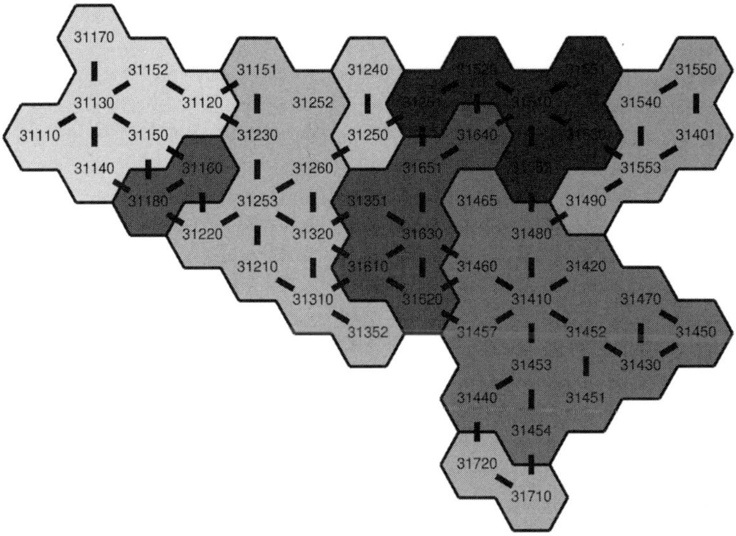

Fig. 12.3. Political suggestion; $b_{\max} = 2.63$ €

$$b_{\text{income}} = \sum_{u,v \in V} W_{uv}(z_{uv} - d_{uv}) \tag{12.2}$$

as the deviation in the income of the public transportation company. This expression assumes that customers do not change their behavior because of tariff changes. Since this is not realistic in practice, we add *price elasticity factors* when evaluating (12.2) for getting a better approximation of the income. There are different definitions of the price elasticity; the one we use is the following. Let d_{uv} denote the currently existing price for traveling from u to v. Then the price elasticity factor p_{elast} is given by

$$p_{elast} = \frac{\frac{W_{uv}^{new} - W_{uv}}{W_{uv}}}{\frac{z_{uv} - d_{uv}}{d_{uv}}}, \tag{12.3}$$

where the number of new customers W_{uv}^{new} is the only unknown value in this formula. Solving this equation for W_{uv}^{new} and replacing W_{uv} by W_{uv}^{new} in (12.2) yields a more realistic approximation of the new income.

Formula (12.3) can be found , e.g., in [Höh77], and sometimes is referred to as the *shrinkage ratio*, compare, e.g., [LMM81]. Approximately, formula (12.3) means that increasing the ticket price by 1% will result in a change in the number of customers of p_{elast}%. Since increasing the price usually yields a reduction in the number of customers, and a reduction of the prices increases the number of customers, the price elasticity factors p_{elast} are negative in

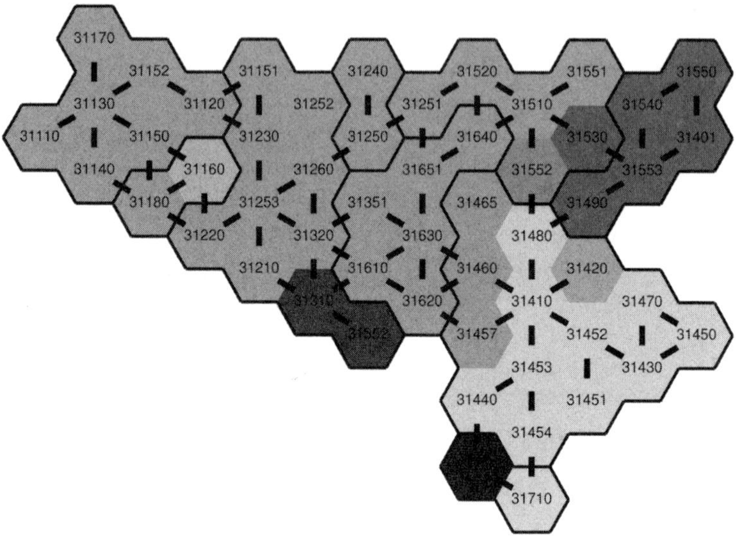

Fig. 12.4. Solution of single linkage; $b_{\max} = 2.56$ €

most cases. The only exceptions are prestige goods, which are bought because they are expensive. Furthermore, if $p_{elast} < -1$, the demand is called *elastic*, while an elasticity factor $0 > p_{elast} > -1$ describes an inelastic demand. If $p_{elast} = -1$, the income will not change, i.e., the increase of the price and the reduction of customers cancels out.

In (local) public transportation, all studies we are aware of confirm that the demand is inelastic, i.e., that

$$-1 < p_{elast} < 0.$$

The particular elasticity factor which is widely used is based on a rule of thumb, derived from the classical Simpson and Curtin formula, see [Cur68]. They claimed a factor of −0.3. Many other — more recent — studies [LMM81, ta98, HK98a] roughly confirm this factor for single trip tickets, but suggest a smaller factor, e.g., −0.2 for weekly, monthly and annual tickets. In practical reports, often a more pessimistic point of view is used. Namely, the price elasticity factor is only used for relations (u, v) for which the tickets get more expensive, while for relations with a reduction in ticket price it is assumed that the demand does not change. Neglecting new customers which might be attracted in such relations makes sure that the new income is treated pessimistically, i.e., it will not happen, that the loss in income is higher than estimated.

Moreover, the public transportation companies are interested in learning about which customers are hardly affected and which customers will be really

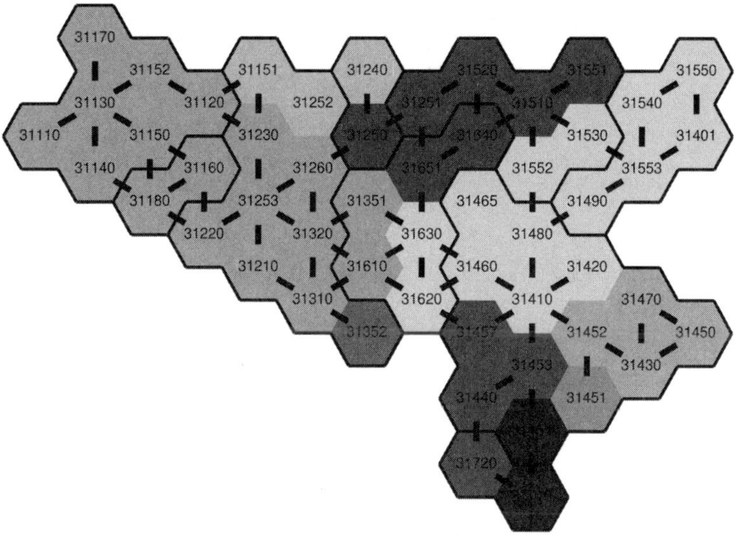

Fig. 12.5. Solution of greedy algorithm; $b_{\max} = 1.92$ €

annoyed. To this end, WabPlan presents graphics (for each ticket category separately) showing all trips for which the fare will increase or decrease dramatically (see Figure 12.6 as an example). Statistical evaluations of the number of relations with price changes, the number of customers affected, and the absolute and relative amount of the price changes are also computed.

For practical purposes a lot of special rules for using fare zones are common. A lot of them have also been implemented in our algorithms and tested within our projects.

Empty zones: In most zone tariff systems, *empty zones* are used to increase the number of zones in a journey and hence the fare for special relations. Note that most given tariff systems can be modeled as a counting zone tariff system, if enough empty zones are allowed.

Border stations: To avoid injustice, stations can be located on zone borders, meaning that they belong to more than one zone. Since the zone tariff system should be clear and understandable, we usually try to avoid this. Fortunately, in many cases it turned out that border stations can be avoided without losing anything in the objective values only by changing the zone design.

Special rules for large zones: Also, some zones might be so large that they have to be counted twice when passing them. Moreover, a special fare structure can be implemented within large zones.

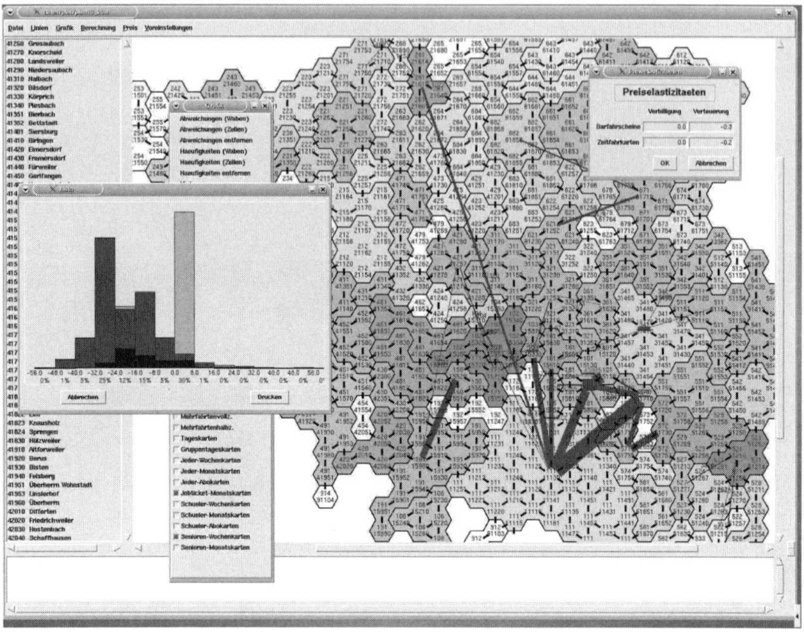

Fig. 12.6. Zone planning with the software *WabPlan* in the state Saarland. The lines show the (fictional) customers who will have a change in their ticket price which is more than 5 Percent.

Sometimes, zone tariff systems are used to give discounts or to make public transportation more convenient only for a special group of customers. In the state of Sachsen-Anhalt, e.g., a zone tariff system was implemented only for customers using at least two different public transportation companies. The goal was to make these trips more attractive by

- offering a good ticket price, and
- providing a combination ticket which can be used for all public transportation companies of the trip.

The new ticket prices in such a system should not be cheaper than the current prices of each single public transportation company; otherwise customers could pretend to change into another bus or train to be allowed to pay the cheaper price. Since customers who wish to change are still allowed to buy separate tickets instead of the new combination ticket, they will always choose the cheaper possibility. Hence, no customer will have to pay more as before, so it is not possible to achieve a gain in the income of the public transportation companies. The resulting zone design problem should hence be as attractive as possible, but with only a small loss in the income of the public transportation companies.

A

Integer Programming

Here we collect some fundamental results of integer programming which are used throughout the text. We assume that the reader is familiar with linear programming, and recommend textbooks as [NW88, Wol98] for the theory of integer programming. All results of this short appendix can be found in both of these books.

We consider the following integer programming problem:

$$
\textbf{(IP)} \quad \begin{array}{ll} \min & cx \\ \text{s.t.} & Ax \le b \\ & x \in \mathbb{Z}^n, \end{array}
$$

where $c \in \mathbb{R}^n, b \in \mathbb{R}^m$ and A is an $m \times n$ matrix.

Definition A.1. *The* **linear programming relaxation (LP-IP)** *or short* **LP-relaxation** *of an integer program (IP) is given by deleting the integer constraints, i.e.*

$$
\begin{array}{ll} \textbf{(IP)} & \min\{cx \text{ s.t. } Ax \le b, x \in \mathbb{Z}^n\} \\ \textbf{(LP-IP)} & \min\{cx \text{ s.t. } Ax \le b, x \in \mathbb{R}^n\}. \end{array}
$$

The relation between an integer program (IP) and its LP-relaxation (LP-IP) is the following.

Lemma A.2. *Let*

$$
\begin{aligned} z^* &= cx^* = \min\{cx \text{ s.t. } Ax \le b, x \in \mathbb{Z}^n\} \\ \bar{z} &= c\bar{x} = \min\{cx \text{ s.t. } Ax \le b, x \in \mathbb{R}^n\}. \end{aligned}
$$

Then

(i) $\bar{z} \le z^*$
(ii) If $\bar{x} \in \mathbb{Z}^n$ *then* $z^* = \bar{z}$ *and* $x^* = \bar{x}$ *is an optimal solution of (IP).*

The lemma states that each solution of the LP-relaxation gives a lower bound on the original program. Moreover, if the solution of the LP-relaxation happens to be integer, it is the optimal solution of the integer program. This observation can be used to find polynomially solvable special cases of integer programs, namely, if all vertices of the feasible set $\{x : Ax \leq b, x \in \mathbb{R}^n\}$ of (LP-IP) are integer. In this case, we know that all basic solutions are integer, and hence we conclude from the theory of linear programming that there exists an *integer* optimal solution $\bar{x}$. Lemma A.2 then yields the optimality of $\bar{x}$ for the original integer program. Thus, in that case it is enough to solve the LP-relaxation of the integer program.

To identify integer programs in which all vertices of the feasible set are integer, we need the concept of *total unimodularity.*

Definition A.3. *A matrix A is* **totally unimodular**, *if $det(B) \in \{-1, 0, 1\}$ for all square submatrices B of A.*

If A is a totally unimodular matrix, then

- $a_{ij} \in \{-1, 0, 1\}$ for all entries of the matrix A,
- the transposed A^T is totally unimodular,
- $(A|I)$ is totally unimodular (I is the $m \times m$ unit matrix).

An example of a matrix with entries only in $\{-1, 0, 1\}$, but that is not totally unimodular, is the following:

$$\begin{pmatrix} 1 & 0 & -1 \\ 0 & -1 & 0 \\ 1 & 1 & 1 \end{pmatrix}.$$

Totally unimodular matrices are important since their corresponding integer programs can be solved by linear programming methods.

Theorem A.4. *Let $b \in \mathbb{Z}^m$ and let A be totally unimodular. Then all vertices of $\{x \in \mathbb{R}^n : Ax \leq b\}$ are integer. Consequently, each optimal basic solution of (LP-IP) is an optimal solution for (IP).*

Examples of totally unimodular matrices (resulting in polynomially solvable integer programs) are

- incidence matrices of networks; this result is used in Sections 7.1 and 8.4 in Part II,
- matrices with the consecutive ones property, which play an important role in Sections 3.4, 3.6, and in 4.3 in Part I,
- network matrices, mentioned both in Section 3.4 (Part I) and in Section 7.3 (Part II).

B

Bicriteria Optimization

Here we briefly introduce bicriteria optimization problems, which are discussed in Chapters 4 and 9. For an introduction to multicriteria optimization we refer the reader to textbooks (e.g. [Ehr00]), a detailed state-of-the art survey of the field is given by [FGE05]. The results of this short overview can be found e.g. in [Ehr00].

All bicriteria problems treated in this text are combinatorial optimization problems, such that we will assume that the feasible set Feas consists of a discrete set of points, i.e.,

$$\text{Feas} \in \mathbb{Z}^n.$$

The bicriteria optimization problem we consider is given by a feasible set $\text{Feas} \subseteq \mathbb{Z}^n$ and two objective functions $f_1, f_2 : \text{Feas} \to \mathbb{R}$.

$$\textbf{(BP)} \quad \min_{x \in \text{Feas}} \begin{pmatrix} f_1(x) \\ f_2(x) \end{pmatrix}.$$

Definition B.1. *Let* $x_1, x_2 \in \text{Feas}$.

- x_1 **dominates** x_2 *if*

$$f_1(x_1) \leq f_1(x_2) \text{ and}$$
$$f_1(x_1) \leq f_1(x_2),$$

  where at least one of the inequalities is strict.
- $x \in \text{Feas}$ *is a* **Pareto solution**, *if there does not exist any* $y \in \text{Feas}$ *that dominates* x.

The goal in bicriteria optimization is to determine the Pareto solutions, i.e., the set of all $x \in \text{Feas}$ which are non-dominated. However, it often is enough to know the objective values of the Pareto solutions. To this end, let

$$f(\text{Feas}) = \left\{ \begin{pmatrix} f_1(x) \\ f_2(x) \end{pmatrix} : x \in \text{Feas} \right\}$$

denote the objective space of (BP). Then a point $\begin{pmatrix} f_1(x) \\ f_2(x) \end{pmatrix} \in f(\text{Feas})$ is called *efficient*, if $x \in \text{Feas}$ is a Pareto solution.

For an illustration, see Figure B.1. In this example let us assume that the set of objective values for all points $x \in \text{Feas}$ is given by the points depicted in the figure. Then the five filled points $p_1, \ldots, p_5$ are not dominated by any other point, i.e., exactly these points are efficient.

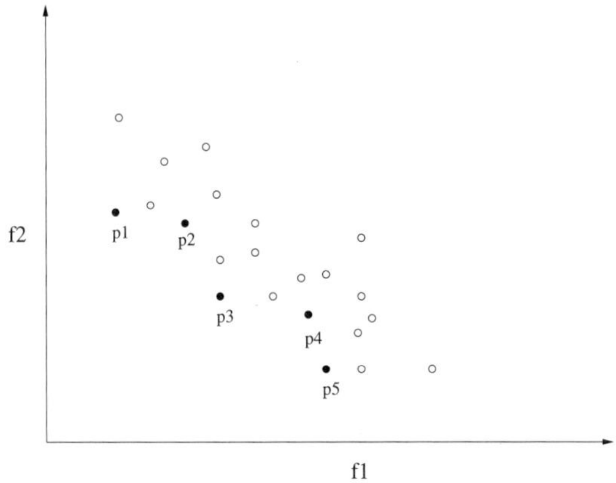

Fig. B.1. Efficient solutions of a bicriteria optimization problem.

For finding Pareto solutions we can solve a one-criteria optimization problem. This method is called *weighted sum scalarization.*

Theorem B.2. *If x is an optimal solution of*

$$\textbf{BP}(\lambda) \qquad \min_{x \in \text{Feas}} \lambda f_1(x) + (1 - \lambda) f_2(x)$$

for some $0 < \lambda < 1$, then x is a Pareto solution of (BP).

Unfortunately, not all Pareto solutions can be found by weighted sum scalarization, if the set $\text{Feas} \subseteq \mathbb{Z}^n$ consists of a discrete set of points. In Figure B.1, the efficient points p_1, p_3, and p_5 can be found by solving a weighted sum problem, while no λ exists such that p_2 and p_4 are optimal solutions of $\text{BP}(\lambda)$.

Definition B.3. *A Pareto solution x is called* **supported** *if there exists λ with $0 < \lambda < 1$ such that x is the optimal solution of BP(λ).*

Note that the term *supported* is due to the following fact: If x is a supported Pareto solution, then $f(x) = \begin{pmatrix} f_1(x) \\ f_2(x) \end{pmatrix}$ lies on the boundary of the convex hull of $f(\text{Feas})$. Hence there exists a supporting line of $f(\text{Feas})$ passing through $f(x)$.

By weighted sum scalarization, we find exactly the set of supported Pareto solutions. With the following result we can also find non-supported Pareto solutions. It uses the constraint versions of (BP).

Lemma B.4. *Let $\{i, j\} = \{1, 2\}$ and let x be a unique optimal solution of*

$$\min\{f_i(x) : x \in \text{Feas} \quad \textit{and} \quad f_j(x) \le y_j\}.$$

Then x is a Pareto solution of (BP).

Finally, we state the definition of lexicographic minimal solutions.

Definition B.5. *Let $\{i, j\} = \{1, 2\}$. x is called a* **lexicographic minimal** *solution of (BP) for the order (f_i, f_j) if for all $y \in \text{Feas}$ one of the following conditions holds.*

- *Either $f_i(x) < f_i(y)$,*
- *or $f_i(x) = f_i(y)$ and $f_j(x) \le f_j(y)$.*

Note that the lexicographic minimal solutions are always Pareto solutions.

C

Gauges as Distance Measures

In the stop location problem (Part I) it is important to choose a suitable function to measure the distance from a demand point to a stop or station. Hence, in this section we briefly introduce the concept of gauges. For more information, the reader is referred to [Min67], which is also the basis of this appendix. Note that gauges have been used frequently in location theory, see, e.g., [DKSW01]. Although the following definitions and results can directly be transferred to $\mathbb{R}^n$ we only present the notation for two dimensions, since we obviously only deal with stop location in the plane.

Geometrical observations play an important role for defining the candidate set in Part I. Thus, it makes sense to use the following "geometrical" definition of a norm.

Definition C.1. *Let B be a compact convex set in $\mathbb{R}^2$ with nonempty interior which is symmetric with respect to the origin. Let $x \in \mathbb{R}^2$. Then define the* **norm** $\gamma : \mathbb{R}^2 \to \mathbb{R}$ *as*

$$\gamma(x) := \inf\{\lambda > 0 : x \in \lambda B\}.$$

The following Lemma C.2 states that γ satisfies the properties required for norms, i.e., for all $x, y \in \mathbb{R}^2$ and $\lambda \in \mathbb{R}$ we have

$$\gamma(x) \geq 0 \tag{C.1}$$

$$\gamma(x) = 0 \Longleftrightarrow x = 0 \tag{C.2}$$

$$\gamma(\lambda x) = |\lambda|\gamma(x) \text{ and} \tag{C.3}$$

$$\gamma(x + y) \leq \gamma(x) + \gamma(y). \tag{C.4}$$

On the other hand, all norms can be characterized by their **unit balls** B.

Lemma C.2. *The following results hold.*

1. *Let γ be given as in Definition C.1. Then γ satisfies (C.1) – (C.4).*
2. *Let $\gamma : \mathbb{R}^2 \to \mathbb{R}_{\geq 0}$ be given, such that γ satisfies (C.1) – (C.4). Then its unit ball $B_\gamma = \{x \in \mathbb{R}^2 : \gamma(x) \leq 1\}$ is a compact convex set with nonempty interior which is symmetric with respect to the origin.*

Examples for norms are

- the Euclidean norm $l_2(x) = \sqrt{(x_1)^2 + (x_2)^2}$,
- the Manhattan norm $l_1(x) = |x_1| + |x_2|$,
- the maximum norm (or Chebyshev norm) $l_\infty(x) = \max\{|x_1|, |x_2|\}$,
- the p-norms $(1 \leq p < \infty)$ $l_p(x) = \sqrt[p]{|x_1|^p + |x_2|^p}$.

The corresponding unit balls for the first three examples are depicted in Figure C.1.

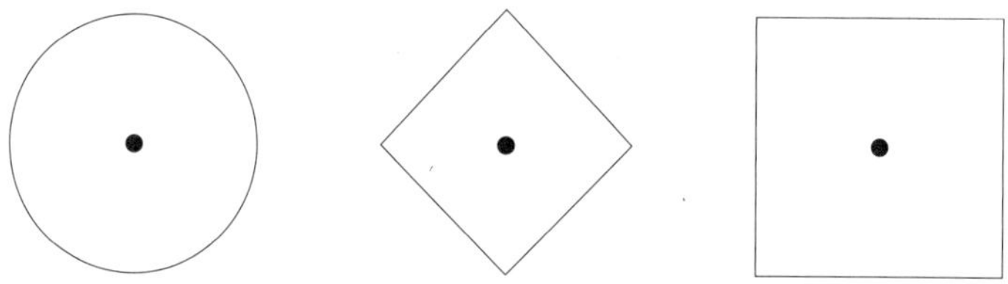

a) Euclidean norm b) Manhattan norm c) maximum norm

Fig. C.1. The unit balls of the Euclidean, the Manhattan and the maximum norm.

If we drop the assumption that the set B is symmetric in Definition C.1, then B does not define a norm, but a *gauge*.

Definition C.3. *Let B be a compact convex set in $\mathbb{R}^2$ containing the origin in its interior. The* **gauge** *of x with respect to B is then defined as*

$$\gamma(x) := \inf\{\lambda > 0 : x \in \lambda B\}.$$

Note that the convexity of B is still required, but $\gamma(-x) = \gamma(x)$ does not hold in general without the symmetry assumption.

The distance between two points $x, y \in \mathbb{R}^2$ is defined by

$$d(x, y) = \gamma(y - x).$$

If γ is a norm we obtain $d(x, y) = d(y, x)$, while this need not be the case for a gauge γ.

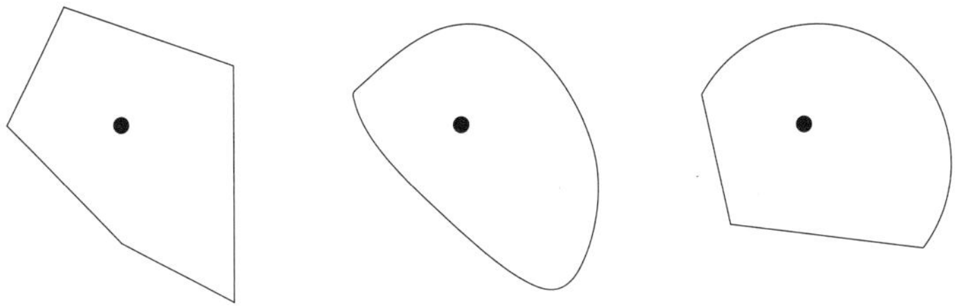

Fig. C.2. Examples of gauges that are not norms.

Examples of gauges which are not norms are given in Figure C.2. In public transportation, a distance measure for some demand point d can be constructed by defining

$$B_d = \{x \in \mathbb{R}^2 : x \text{ can be reached within a time of } r \text{ minutes}\}$$

as the unit ball for demand point d and using the corresponding gauge γ_{B_d} as the required distance measure. Note that the shape of B_d depends on the road network structure and on the public transportation system in the area around d.

4

Frequently Used Notation

General notation:

$\mathbb{N}$	natural numbers including 0
$\mathbb{Z}$	integer numbers
$\mathbb{R}$	real numbers
$\{u, v\}$	undirected edge in a graph
(u, v)	directed edge in a digraph
A^T	transpose of a matrix
Θ	node-arc incidence matrix of a network

Public transportation network:

$\text{PTN} = (V, E)$	public transportation network
V	set of stops of the PTN
	elements denoted by u, v, v_i
E	set of direct rides in the PTN
	elements denoted by $\{u, v\}$, or by e

Data about customers:

$W = (W_{uv})$	OD-matrix, i.e., number of customers traveling from u to v
c_e	traffic load on edge $e \in E$,
	i.e., number of customers traveling along edge e
c_v	traffic load through stop $v \in V$
	i.e., number of customers traveling through stop v

w_d	demand in demand point $d \in \mathcal{D}$
w_D	demand in demand region $D \in \mathcal{D}$
w_p	number of customers traveling along path $p \in \mathcal{P}$
w_a	"traffic load" on activity $a \in \mathcal{A}$
w_i	number of customers really getting off at event $i \in \mathcal{E}$

Stop location:

$\mathcal{D}$	set of demand points or demand regions elements denoted by d for demand points and by D for demand regio
$G = (V, E)$	given graph in which the stops are to be located
$\mathcal{T}$	tracks, given as points on the embedding of graph G
γ_d	gauge distance function of demand point $d \in \mathcal{D}$
γ_D	gauge distance function of demand region $D \in \mathcal{D}$
$g(s)$	edge e or node v in which point $s \in \mathcal{T}$ is located
S^{ex}	existing stops or stations
S	set of new stops or stations

Delay management:
(in the notation using event-activity networks)

$\mathcal{P}$	set of paths of customers
$\mathcal{N} = (\mathcal{E}, \mathcal{A})$	event-activity network
$\mathcal{E}_{arr}$	arrival events
$\mathcal{E}_{dep}$	departure events
$\mathcal{A}_{wait}$	waiting activities
$\mathcal{A}_{drive}$	driving activities
$\mathcal{A}_{change}$	changing activities
Π_i	time for event $i \in \mathcal{E}$ in the original timetable
x_i	time for event $i \in \mathcal{E}$ in the perturbed timetable
y_i	delay of event $i \in \mathcal{E}$
s_a	slack time of activity $a \in \mathcal{A}$
d_i	source delay of event $i \in \mathcal{E}$
$\mathcal{E}_{del}$	set of delayed events

Tariff planning:

$\mathcal{Z}$	zone partition
$c(p)$	price for traveling through $p+1$ zones, i.e., for passing p zone borders
L	required number of zones
n_{uv}	number of zones needed for traveling from u to v
d_{uv}	reference price for a ticket from u to v
M_p	set of relations passing p zones

5

List of the Main Problems

References

[AC97] A. Adamski and W. Chmiel. Optimal service synchronization in public transport. In *Transportation Systems, IFAC/IFORS Symposium, Chania, Greece, June 1997*, pages 1283–1287, 1997.

[Ack99] T. Ackermann. *Die Bewertung der Pünktlichkeit als Qualitätsparameter im Schienenpersonennahverkehr auf Basis der direkten Nutzenmessung.* PhD thesis, Universität Stuttgart, 1999.

[AD96] B. Adenso-Díaz. An SA/TS mixture algorithm for the scheduling tardiness problem. *European Journal of Operational Research*, 88:516–524, 1996.

[ADGGT99] B. Adenso-Díaz, M. Oliva González, and P. González-Torre. On-line timetable re-scheduling in regional train services. *Transportation Research*, 33B:387–398, 1999.

[AFT02] A.Caprara, A. Fischetti, and P. Toth. Modeling and solving the timetabling problem. *Operations Research*, 50(5):851–861, 2002.

[APW02] L. Anderegg, P. Penna, and P. Widmayer. Online train disposition: to wait or not to wait? *Electronic Notes in Theoretical Computer Science*, 66(6), 2002.

[Ass80] A.A. Assad. Models for rail transportation. *Transportation Research A*, 14(3):205–220, 1980.

[Bau05] I. Bauerdorf. Sollen Anschlussverbindungen bei Verspätungen unterbrochen werden? Ein Ansatz zur Formulierung der Fragestellung in der Theorie des option pricing. In *proceedings of Operations Research 2005*, 2005.

[BBH90] P. Brucker, R.E. Burkard, and J. Hurink. Cyclic schedules for r irregularly ocurring events. *Journal of Computational and Applied Mathematics*, 30:173–189, 1990.

[BD92] J.H. Bookbinder and A. Desilets. Transfer optimization in a transit network. *Transportation Science*, 26(2):106–118, 1992.

[Bel58] R. Bellman. On a routing problem. *Quarterly Applied Mathematics*, 16:87–90, 1958.

[BGJ+05] N. Bissantz, S. Güttler, J. Jacobs, S. Kurby, T. Schaer, A. Schöbel, and S. Scholl. DisKon - Disposition und Konfliktlösungs-management für die beste Bahn. *Eisenbahntechnische Rundschau (ETR)*, 45(12):809–821, 2005. (in German).

[BGL98] R. Borndörfer, M. Grötschel, and A. Löbel. Optimization of transportation systems. In *Acta Forum Engelberg 98*. VDF Hochschulverlag an der ETH Zürich, 1998.

[BGP04a] R. Borndörfer, M. Grötschel, and M. E. Pfetsch. Models for line planning in public transport. Technical Report 04-10, ZIP Berlin, 2004.

[BGP04b] R. Borndörfer, M. Grötschel, and M. E. Pfetsch. A path-based model for line planning in public transport. Technical Report 05-18, ZIP Berlin, 2004.

[BH86] P. Brucker and J. Hurink. A railway scheduling problem. *Zeitschrift für Operations Research*, 30:223–227, 1986.

[BHK99] P. Brucker, S. Heitmann, and S. Knust. Scheduling railway traffic at a construction site. In K.H. Kim (eds.): in: H.-O. Günther, editor, *Container Terminals and Automated Transport Systems Logistics Control Issues and Quantitative Decision Support*. Springer, 1999.

[BK03] L. Babel and H. Kellerer. Design of tariff zones in public transportation systems: Theoretical results and heuristics. *Mathematical Methods of Operations Research*, pages 358–374, 2003.

[BKS92] W. Bein, J. Kamburowski, and M. Stallmann. Optimal reduction of two-terminal directed acyclic graphs. *SIAM Journal on Computing*, 21(6):1112–1129, 1992.

[BKZ96] M.R. Bussieck, P. Kreuzer, and U.T. Zimmermann. Optimal lines for railway systems. *European Journal of Operational Research*, 96(1):54–63, 1996.

[BLNN98] U. Brännlund, P.O. Lindberg, A. Nou, and J.-E. Nilsson. Railway timetable using lagrangian relaxation. *Transportation Science*, 32(4):358–369, November 1998.

[BM95] M.H. Baaj and H.S. Mahmassani. Hybrid route generation heuristic algorithm for the design of transit networks. *Transportation Research C*, 3:31–50, 1995.

[BNP05] R. Borndörfer, M. Neumann, and M. E. Pfetsch. Fare planning for public transport. Technical Report 05-20, ZIB, Berlin, 2005.

[BT81] L. A. Bowman and M.A. Turnquist. Service frequency, schedule reliability, and passengers waiting times at transit stops. *Transportation research*, 15A(6), 1981.

[BT96] L. Bianco and P. Toth, editors. *Advanced methods in transportation analysis*. Transportation Analysis. Springer, 1996.

[Bur86] R.E. Burkard. Optimal schedules for periodically recurring events. *Discrete Applied Mathematics*, 15:167–180, 1986.

[Bus97] M.R. Bussieck. *Optimal lines in public transport*. PhD thesis, Technische Universität Braunschweig, 1997.

[BWZ97] M.R. Bussieck, T. Winter, and U. Zimmermann. Discrete optimization in public rail transport. *Mathematical Programming*, 79:415–444, 1997.

[Car98] M. Carey. Optimizing scheduled times, allowing for behavioural response. *Transportation Research*, 32B:329–342, 1998.

[Car99] M. Carey. Ex ante heuristic measures of schedule reliability. *Transportation Research*, 53B(3):473–494, 1999.

[CFT99] A. Caprara, M. Fischetti, and P. Toth. A heuristic method for the set covering problem. *Operations Research*, 47(5):730–743, 1999.

[CG02] S. Carrese and S. Gori. An urban bus network design procedure. In M. Patriksson and M. Labbé, editors, *Transportation Planning. State of the Art*, volume 64 of *Applied Optimization*, chapter 11. Kluwer, 2002.

[CH90] B. Chen and P.T. Harker. Two moments estimation of the delay on single-track rail lines with scheduled traffic. *Transportation Science*, 24(4):261–275, November 1990.

[CM82] J.C.M. Climaco and E.Q.V. Martins. A bicriterion shortest path algorithm. *European Journal of Operational Research*, 11:399–404, 1982.

[Con02] Nicolas Condette. Set covering problems with consecutive ones property. Master's thesis, University of Kaiserslautern, 2002.

[CR74] R. Church and C. ReVelle. The maximal covering location problem. *Papers of the Regional Science Association*, 32:101–118, 1974.

[CTV98] J.F. Cordeau, P. Toth, and D. Vigo. A survey of optimization models for train routing and scheduling. *Transportation Science*, 1998.

[Cur68] J. F. Curtin. Effects of fares on transit riding. *Highway Research Board*, 213, 1968.

[CvDZ96] M.T. Claessens, N.M. van Dijk, and P.J. Zwaneveld. Cost optimal allocation of rail passenger lines. *European Journal on Operational Research*, 110, 1996.

[CW86] A. Ceder and N.H.M. Wilson. Bus network design. *Transportation Research B*, 20:331–344, 1986.

[DAL82] M. J. Demetsky, M. Asce, and B. B.-M. Lin. Bus stop location and design. *Transportation Engineering Journal*, 108:313–327, 1982.

[DBP95] J.R. Daduna, I. Branco, and J.M.P. Paixao, editors. *Computer-Aided Transit Scheduling*, volume 430 of *Lecture Notes in Economics and Mathematical systems*. Springer, 1995.

[DDGW97] P. De, J. Dunne, J. Gosh, and C. Wells. Complexity of the discrete time-cost tradeoff problem for project networks. *Operations Research*, 45(2):302–306, 1997.

[DHE96] E. Demeulemeester, W. Herroelen, and S. Elmaghraby. Optimal procedures for the discrete time/cost trade-off problem in project networks. *European Journal of Operational Research*, 88:50–68, 1996.

[Die78] H. Dienst. *Linienplanung im spurgeführten Personenverkehr mit Hilfe eines heuristischen Verfahrens*. PhD thesis, Technische Universität Braunschweig, 1978.

[Dij59] E. Dijkstra. A note on two problems in connexion with graphs. *Numerische Mathematik*, 1:269–271, 1959.

[DKSW01] Z. Drezner, K. Klamroth, A. Schöbel, and G. Wesolowsky. The weber problem. In Z. Drezner and H.W. Hamacher, editors, *Location Theory - Applications and Theory*, chapter 1, pages 1–36. Springer, 2001.

[DO74] B.S. Duran and P.L. Odell. *Cluster Analysis — A Survey*. Lecture Notes in Economics and Mathematical Systems 100. Springer, Berlin-Heidelberg-New York, 1974.

[Dom89] W. Domschke. Schedule synchronisation for public transit networks. *OR Spektrum*, 11:17–24, 1989.

[DR92] J.R. Daduna and J.-M. Rousseau, editors. *Computer-Aided Transit Scheduling*, volume 386 of *Lecture Notes in Economics and Mathematical systems*. Springer, 1992.

[DRF+98] E. Demeulemeester, B. De Reyck, B. Foubert, W. Herroelen, and M. Vanhoucke. New computational results on the discrete time/cost trade-off problem in project networks. *Journal of the Operational Research Society*, 49:1153–1163, 1998.

[DV95] J.R. Daduna and S. Voß. Practical experiences in schedule synchronization. In *Computer-Aided Transit Scheduling*, volume 430 of *Lecture Notes in Economics and Mathematical Systems*, pages 39–55. Springer, 1995.

[DW88] J.R. Daduna and A. Wren, editors. *Computer-Aided Transit Scheduling*, volume 308 of *Lecture Notes in Economics and Mathematical systems*. Springer, 1988.

[EF02] O. Engelhardt-Funke. *Stochastische Modellierung und Simulation von Verspätungen in Verkehrsnetzen für die Anwendung der Fahrplanoptimierung*. PhD thesis, Univerität Clausthal, 2002.

[EFK01a] O. Engelhardt-Funke and M. Kolonko. Cost-benefit analysis of investments into railway networks with randomly pertubed operations. In S. Voß and J. Daduna, editors, *Computer-Aided Transit Scheduling*, volume 505 of *Lecture Notes in Economics and Mathematical systems*, pages 442–459. Springer, 2001.

[EFK01b] O. Engelhardt-Funke and M. Kolonko. Simulating delays for realistic timetable-optimization. In *Operations Research Proceedings 2001*, pages 9–15. Springer, 2001.

[EG02] M. Ehrgott and X. Gandibleux. Multiobjective combinatorial optimization. In M. Ehrgott and X. Gandibleux, editors, *Multiple Criteria Optimization. State of the Art. Annotated Bibliographic Surveys*, pages 369–444. Kluwer, 2002.

[EH72] J. Elzinga and D.W. Hearn. Geometrical solutions for some minimax location problems. *Transportation Science*, 4:379–394, 1972.

[Ehr00] M. Ehrgott. *Multiple Criteria Optimization*, volume 491 of *Lecture Notes in Economics and Mathematical Systems*. Springer, Berlin, 2000.

[Elm77] S.E. Elmaghraby. *Activity Networks*. Wiley Interscience Publication, 1977.

[Fay00] A. Fay. A fuzzy knowledge-based system for railway traffic control. *Engineering applications of Artiicial intelligence*, 13:719–729, 2000.

[FF62] L. R. Ford and D. R. Fulkerson. *Flows in Networks*. Princeton University Press, 1962.

[FGE05] J. Figueira, S. Greco, and M. Ehrgott, editors. *Multiple Criteria Decision Analysis: State of the Art Surveys*, volume 78 of *International Series in Operations Research & Management Science*. Springer, New York, 2005.

[Fle91] B. Fleischmann. Synchronization of transfers in public transit networks with cyclic schedules. Technical report, Institut für Unternehmensforschung, Universität Hamburg, 1991.

[Flo62] R.W. Floyd. Algorithm 97: Shortest path. *Communications of the ACM*, 5(6):345, 1962.

[GBO99] R. Goverde, P Bovy, and G. Olsder. The max-plus algebra approach to transportation problems. In H. Meersman, editor, *Transport Modelling/Assessment,*, volume 3 of *World Transport Research*, pages 377–391. 1999.

[Geo68] A. M. Geoffrion. Proper efficiency and the theory of vector maximization. *Journal of Mathematical Analysis and Applications*, 22:618–630, 1968.

[Ger04] B. Gerards, editor. *Algorithmic Methods and Models for Optimization of Railways (ATMOS) 2003*, volume 92 of *Electronic Notes in Theoretical Computer Science*, 2004.

[GGJ+04] M. Gatto, B. Glaus, R. Jacob, L. Peeters, and P. Widmayer. Railway delay management: Exploring its algorithmic complexity. In *Proceedings 9th Scandinavian Workshop on Algorithm Theory (SWAT)*, volume 3111 of *LNCS*, pages 199–211, 2004.

[GHL06] L. Giovanni, G. Heilporn, and M. Labbé. Optimization models for the delay management problem in public transportation. *European Journal of Operational Research*, 2006. to appear.

[Gin01] A. Ginkel. Event-activity networks in delay management. Master's thesis, Universität Kaiserslautern, 2001.

[GJ79a] M.R. Garey and D.S. Johnson. *Computers and Intractability — A Guide to the Theory of NP-Completeness.* Freeman, San Francisco, 1979.

[GJ79b] M.R. Garey and D.S. Johnson. *Computers and Intractability — A Guide to the Theory of NP-Completeness.* H.W. Freeman and Company, San Francisco, 1979.

[GJP+04] M. Gatto, R. Jacob, L. Peeters, B. Weber, and P. Widmayer. Theory on the tracks: A selection of railway optimization problems (column: Algorithmics). In *Bulletin of the EATCS*, volume 84, pages 41–70, 2004.

[GJPS05] M. Gatto, R. Jacob, L. Peeters, and A. Schöbel. The computational complexity of delay management. In D. Kratsch, editor, *Graph-Theoretic Concepts in Computer Science: 31st International Workshop (WG 2005)*, volume 3787 of *Lecture Notes in Computer Science*, 2005.

[GJPW06] M. Gatto, R. Jacob, L. Peeters, and P. Widmayer. On-line delay management on a single train line. In *Algorithmic Methods for Railway Optimization*, Lecture Notes in Computer Science. Springer, 2006. presented at ATMOS 2004, to appear.

[GKS+06] F. Geraets, L. Kroon, A. Schöbel, D. Wagner, and C. Zaroliagis, editors. *Algorithmic Methods for Railway Optimization.* Lecture Notes in Computer Science. Springer, 2006. to appear.

[Gle75] J. Gleason. A set covering approach to bus stop allocation. *Omega*, 3:605–608, 1975.

[GM06] D. Gattuso and G. Musolino. A simulation approach of fare integration in regional transit services. In *Algorithmic Methods for Railway Optimization*, Lecture Notes in Computer Science. Springer, 2006. presented at ATMOS 2004, to appear.

[Gol89] D.E. Goldberg. *Genetic Algorithms in Search, Optimization and Machine Learning.* Addison-Wesley, 1989.

[Goo04] J. Goossens. *Models and algorithms for railway line planning problems.* PhD thesis, University of Maastricht, 2004.

[Gov98a] R.M.P. Goverde. The max-plus algebra approach to railway timetable design. In *Computers in Railways VI: Proceedings of the 6th international conference on computer aided design, manufacture and operations in the railway and other advanced mass transit systems, Lisbon, 1998*, pages 339–350, 1998.

[Gov98b] R.M.P. Goverde. Optimal transfer times in railway timetables. Paper presented at the 6th meeting of the EURO Working Group on Transportation, Götheburg, Schweden, 1998.

[GS02] A. Ginkel and A. Schöbel. The bicriterial delay management problem. Technical report, Universität Kaiserslautern, 2002.

[GvHK02] J. Goossens, C.P.M. van Hoesel, and L.G. Kroon. On solving multi-type line planning problems. Technical Report RM/02/009, University of Maastricht, 2002. METEOR Research Memorandum.

[GvHK04] J. Goossens, C.P.M. van Hoesel, and L.G. Kroon. A branch and cut approach for solving line planning problems. *Transportation Science*, 38:379–393, 2004.

[Gün85] R. Günther. *Untersuchung planerischer und betrieblicher Maßnahmen zur Verbesserung der Anschlußsicherung in städtischen Busnetzen.* PhD thesis, Technische Universität Berlin, 1985.

[Ham95] H.W. Hamacher. *Mathematische Lösungsverfahren für planare Standortprobleme (Mathematical Solution Algorithms for Planar Location Problems).* Vieweg, Braunschweig, 1995.

[Han79] P. Hansen. Bicriterion path problems. In *Multiple Criteria Decision Making. Theory and Applications*, volume 177 of *Lecture Notes in Economics and Mathematical Systems*, pages 109–127. Springer, 1979.

[Hay81] W.L. Hays. *Statistics.* Holt, Rinehart and Winston, New York, 3. edition, 1981.

[HC83] Y.Y. Haimes and V. Chankong. *Multiobjective Decision Making — Theory and Methodology.* North Holland, New York, 1983.

[HdV01] B. Heidergott and R. de Vries. Towards a control theory for transportation networks. *Discrete Event Dynamic Systems*, 11:371–398, 2001.

[Hen85] M.I. Henig. The shortest path problem with two objective functions. *European Journal of Operational Research*, 25:281–291, 1985.

[HH87] S. Holz and R. Hüttmann. Optimierung von Anschlußzeiten. *Der Nahverkehr*, 4, 1987.

[Hig97] A. Higgins. Optimization of train schedules to minimize transit time amd maximize reliability. *Transportation Research*, 31A, 1997.

[HK81] C. Hendrickson and G. Kocur. Schedule delay and departure time decision in a deterministic model. *Transportation Science*, 15:62–77, 1981.

[HK98a] D.A. Hensher and J. King. Establishing fare elasticity regimes for urban passenger transport. Technical Report C37, The University of Sydney, 1998.

[HK98b] A. Higgins and E. Kozan. Modeling train delays in urban networks. *Transportation Science*, 32(4):346–357, November 1998.

[HK01] H.W. Hamacher and K. Klamroth. *Linear and network optimization - a bilingual textbook.* Mathematics International. Vieweg, 2001.

[HKF96] A. Higgins, A. Kozan, and L. Ferreira. Optimal scheduling of trains on a single line track. *Transportation Research*, 30B:147–161, 1996.

[HL05] D. Hochbaum and A. Levin. Optimizing over consecutive 1's and circlular 1's constraints. Technical report, University of Berkeley, 2005.

[HLS+01] H.W. Hamacher, A. Liebers, A. Schöbel, D. Wagner, and F. Wagner. Locating new stops in a railway network. *Electronic Notes in Theoretical Computer Science*, 50(1), 2001.

[HS95] H.W. Hamacher and A. Schöbel. On fair zone design in public transportation. In J.R. Daduna, I. Branco, and J.M.P. Paixao, editors, *Computer-Aided Transit Scheduling*, number 430 in Lecture Notes in Economics and Mathematical Systems, pages 8–22. Springer, Berlin, Heidelberg, 1995.

[HS04] H.W. Hamacher and A. Schöbel. Design of Zone Tariff Systems in Public Transportation. *Operations Research*, 52(6):897–908, 2004.

[Höh77] G.J. Höhn. Die Preiselastizität der Nachfrage — graue Theorie oder handfeste Wirklichkeit für den ÖPNV? *Nahverkehrs Praxis*, 1977.

[Jac04] J. Jacobs. Reducing delays by means of computer-aided 'on-the-spot' rescheduling. In *Computers in Railways (Comprail) IX*, pages 603–612, 2004.

[Joh82] D. S. Johnson. The NP-completeness column: An ongoing guide. *Journal of Algorithms*, 3:182–195, 1982.

[Kli00a] N. Kliewer. Mathematische Optimierung zur Unterstützung kundenorientierter Disposition im Schienenverkehr. Master's thesis, Universität Paderborn, 2000.

[Kli00b] H. Klingele. Verfahren zur Optimierung eines Linienkonzeptes der Deutschen Bahn AG. Master's thesis, Universität Karlsruhe, 2000.

[KNV96] M. Kolonko, K. Nachtigall, and S. Voget. Optimierung von integralen Taktfahrplänen mit genetischen Algorithmen. Technical Report 8/96, Hildesheimer Informatik-Berichte, 1996.

[KPS+03] E. Kranakis, P. Penna, K. Schlude, D.S. Taylor, and P. Widmayer. Improving customer proximity to railway stations. In *Proceedings of the 5th conference on algorithms and complexity*, number 2653 in Lecture Notes in Computer Science, 2003.

[Kri96] M. Krista. *Verfahren zur Fahrplanoptimierung dargestellt am Beispiel der Synchronzeiten.* PhD thesis, Technische Universitat Braunschweig, Fachbereich für Bauingenieur - und Vermessungswesen, 1996.

[Kro97] L. G. Kroon. Routing trains through railway stations: complexity issues. *European Journal of Operational Research*, 98:485–498, 1997.

[KS87] W.D. Klemt and W. Stemme. Schedule synchronization for public transport networks. In *Computer-Aided Transit Scheduling*, volume 308 of *Lecture Notes in Economics and Mathematical Systems*, pages 327–335. Springer, 1987.

[Lie01] A. Liebers. *Analyzing Train Time Table Graphs.* PhD thesis, Universität Konstanz, 2001.

[Lie03] C. Liebchen. Finding short integral cycle bases for cyclic timetabling. In *Proceedings of European Symposium on Algorithms (ESA) 2003*, pages 715–726, 2003.

[LM02] C. Liebchen and R. Möhring. A case study in periodic timetabling. *Electronic Notes in Theoretical Computer Science*, 66(6), 2002.

[LMM81] A. M. Lago, P.D. Mayworm, and J.M. McEnroe. Transit ridership responsiveness to fare changes. *Traffic Quarterly*, 35(1):117–142, 1981.

[LMMO06] G. Laporte, A. Marin, J.A. Mesa, and F.A. Ortega. An integrated methodology for rapid transit network design. In *Algorithmic Methods for Railway Optimization*, Lecture Notes in Computer Science. Springer, 2006. presented at ATMOS 2004, to appear.

[LMO02] G. Laporte, J.A. Mesa, and F.A. Ortega. Locating stations on rapid transit lines. *Computers and Operations Research*, 29:741–759, 2002.

[LMO05] G. Laporte, J.A. Mesa, and F.A. Ortega. Maximizing trip coverage in the location of a single rapid transit alignment. *Annals of Operations Research*, 136:49–63, 2005.

[LMW88] R.F. Love, J.G. Morris, and G.O. Wesolowsky. *Facilities Location*, chapter 3.3, pages 51–60. North-Holland, Amsterdam, 1988.

[LPW04] C. Liebchen, M. Proksch, and F. Wagner. Performance of algorithms for periodic timetable optimization. In *Proceedings of 9th meeting on Computer-Aided Scheduling of Public Transport(CASPT 2004)*, 2004.

[LR05] C. Liebchen and R. Rizzi. A greedy approach to compute a minimum cycle basis of a directed graph. *Information Processing Letters*, 94(3):107–112, 2005.

[MDSF98] A. Murray, R. Davis, R.J. Stimson, and L. Ferreira. Public transportation access. *Transportation Research D*, 3(5):319–328, 1998.

[Mec03] S. Mecke. Standortplanung von Bahnhöfen. Master's thesis, University of Konstanz, 2003.

[Meg83] N. Megiddo. Linear time algorithms for linear programming in $\mathbb{R}^3$ and related problems. *SIAM Journal on Computing*, 12:759–776, 1983.

[Min67] H. Minkowski. *Gesammelte Abhandlungen, Band 2*. Chelsea Publishing Company, New York, 1967.

[MM98] H. Müller-Merbach. *Operations Research*. Vahlens Handbücher der Wirtschafts- und Sozialwissenschaften. Verlag Vahlen, 1998.

[MMO91] J. Mote, I. Murthy, and D.L. Olson. A parametric approach to solving bicriterion shortest path problems. *European Journal of Operational Research*, 53:81–92, 1991.

[MMW04] M.F. Mammana, S. Mecke, and D. Wagner. The station location problem on two intersecting lines. *Electronic Notes in Theoretical Computer Science*, 92:52–64, 2004.

[MSW05] S. Mecke, A. Schöbel, and D. Wagner. Stop location - complexity and approximation issues. In *Proceedings of ATMOS 2005*, 2005.

[Mur01a] A. Murray. Coverage models for improving public transit system accessibility and expanding access. Technical report, Department of Geography, Ohio State University, 2001.

[Mur01b] A. Murray. Strategic analysis of public transport coverage. *Socio-Economic Planning Sciences*, 35:175–188, 2001.

[MW04] S. Mecke and D. Wagner. Solving geometric covering problems by data reduction. In *Proceedings of European Symposium on Algorithms (ESA)*, pages 760–771, 2004.

[Nac94] K. Nachtigall. A branch and cut approach for periodic network programming. Technical Report 29/94, Hildesheimer Informatik-Berichte, 1994.

[Nac96] K. Nachtigall. Periodic network optimization with different arc frequencies. *Discrete applied mathematics*, 69:1–17, 1996.

[Nac97] K. Nachtigall. Cutting planes for a polyhedron associated with a periodic network. Technical report, Institut für Flugführung, 1997.

[Nac98] K. Nachtigall. *Periodic Network Optimization and Fixed Interval Timetables*. Deutsches Zentrum für Luft– und Raumfahrt, Institut für Flugführung, Braunschweig, 1998. Habilitationsschrift.

[Neu75] K. Neumann. *Operations Research Verfahren*, volume III. Carl Hanser Verlag, München Wien, 1975.

[NSS01] S. Nickel, A. Schöbel, and T. Sonneborn. Hub location problems in urban traffic networks. In Niittymaki and Pursula, editors, *Mathematical Methods and Optimization in Transportation Systems*, pages 95–107. Kluwer academic Publishers, 2001.

[NV96] K. Nachtigall and S. Voget. A genetic approach to periodic railway synchronization. *Computers Ops. Res.*, 23(5):453–463, 1996.

[NV97] K. Nachtigall and S. Voget. Minimizing waiting times in integrated fixed interval timetables by upgrading railway tracks. *European Journal of Operational Research*, 103:610–627, 1997.

[NW88] G. L. Nemhauser and L. A. Wolsey. *Integer and Combinatorial Optimization*. Wiley, 1988.

[Odi96] M. A. Odijk. A constraint generation algorithm for the construction of periodic railway timetables. *Transportation Research*, 30B:455–464, 1996.

[Pee02] L. Peeters. *Cyclic Railway Timetabling Optimization*. PhD thesis, ERIM, Rotterdam School of Management, 2002.

[Pen97] K. Penner. Fahrpreisgestaltung im öffentlichen Verkehr. Master's thesis, Universität Kaiserslautern, 1997.

[PK01] L. Peeters and L. Kroon. A cycle based optimization model for the cyclic railway timetabling problem. In S. Voß and J. Daduna, editors, *Computer-Aided Transit Scheduling*, volume 505 of *Lecture Notes in Economics and Mathematical systems*, pages 275–296. Springer, 2001.

[PK03] L. Peeters and L. Kroon. A variable trip time model for cyclic railway timetabling. *Transportation Science*, 37(2):198–212, 2003.

[PL02] M. Patriksson and M. Labbé, editors. *Transportation Planning. State of the Art*, volume 64 of *Applied Optimization*. Kluwer, 2002.

[Pla95] F. Plastria. Continuous location problems. In Z. Drezner, editor, *Facility Location: A survey of applications and methods*, chapter 11, pages 225–262. Springer, New York, Inc., 1995.

[PMP04] D. Pacciarelli, A. Mascis, and M. Pranzo. Scheduling models for short-term railway traffic optimization. In *Proceedings of 9th meeting on Computer Aided Scheduling of Public Transport(CASPT 2004)*, 2004.

[PT82] E. R. Petersen and A. J. Taylor. A structured model for rail line simulation and optimization. *Transportation Science*, 16(2):192–206, 1982.

[Roc84] R.T. Rockafellar. *Network Flows and Monotropic Optimization*. John Wiley, New York, 1984.

[Rou85] J.-M. Rousseau, editor. *Computer Scheduling of Public Transport 2*. North-Holland, 1985.

[RS04] N. Ruf and A. Schöbel. Set covering problems with almost consecutive ones property. *Discrete Optimization*, 1(2):215–228, 2004.

[Ruf02] N. Ruf. Locating train stations: Set covering problems with "almost c1p" matrices. Master's thesis, University of Kaiserslautern, 2002.

[SA00] A.J.V. Skriver and K.A. Andersen. A label correcting approach for solving bicriterion shortest path problems. *Computers and Operations Research*, 27(6):507–524, 2000.

[SBK01] L. Suhl, C. Biederbick, and N. Kliewer. Design of customer-oriented dispatching support for railways. In S. Voß and J. Daduna, editors, *Computer-Aided Transit Scheduling*, volume 505 of *Lecture Notes in Economics and Mathematical systems*, pages 365–386. Springer, 2001.

[Sch94a] A. Schöbel. Methoden der kombinatorischen Optimierung in der Tarifplanung im öffentlichen Personennahverkehr. Master's thesis, Universität Kaiserslautern, 1994. (appeared under maiden name, A. Schumacher).

[Sch94b] A. Schöbel. Fair zone design in public transportation networks. In U. Derigs, A. Bachem, and A.Drexl, editors, *Operations Research Proceedings 1994*, pages 191–196, Berlin, 1994. Springer Verlag.

[Sch96] A. Schöbel. Zone planning in public transportation. In L. Bianco and P. Toth, editors, *Advanced Methods in Transportation Analysis*, Transportation Analysis, pages 117–134. Springer Verlag, 1996.

[Sch01a] M. Schmidt. Modelle zur Linienoptimierung im Zugverkehr unter Berücksichtigung der Nachfrage. Master's thesis, Universität Kaiserslautern, 2001.

[Sch01b] S. Scholl. Anschlusssicherung bei Verspätungen im ÖPNV. Master's thesis, Universität Kaiserslautern, 2001.

[Sch01c] A. Schöbel. A model for the delay management problem based on mixed-integer programming. *Electronic Notes in Theoretical Computer Science*, 50(1), 2001.

[Sch05a] R.-S. Schneider. A new customer-oriented cost model for the line planning problem. Master's thesis, Technische Universität Kaiserslautern, 2005.

[Sch05b] S. Scholl. *Customer-oriented line planning*. PhD thesis, Technische Universität Kaiserslautern, 2005.

[Sch05c] A. Schöbel. Locating stops along bus or railway lines — a bicriterial problem. *Annals of Operations Research*, 136:211–227, 2005.

[Sch06] A. Schöbel. Integer programming approaches for solving the delay management problem. In *Algorithmic Methods for Railway Optimization*, Lecture Notes in Computer Science. Springer, 2006. to appear.

[SHLW02] A. Schöbel, H.W. Hamacher, A. Liebers, and D. Wagner. The continuous stop location problem in public transportation. Technical report, Universität Kaiserslautern, 2002. Report in Wirtschaftsmathematik Nr. 81/2001.

[Skr00] A.J.V. Skriver. A classification of bicriteria shortest path (bsp) algorithms. *Asia-Pacific Journal of Operational Research*, 17(2):199–212, 2000.

[SM97] L. Suhl and T. Mellouli. Supporting planning and operation time control in transportation systems. In *Operations Research Proceedings 1996*, pages 374–379. Springer, 1997.

[SM99] L. Suhl and T. Mellouli. Requirements for, and design of, an operations control system for railways. In *Computer-Aided Transit Scheduling*. Springer, 1999.

[SM01] L. Suhl and T. Mellouli. Managing and preventing delays in railway traffic by simulation and optimization. In *Mathematical Methods on Optimization in Transportation Systems*, pages 3–16. Kluwer, 2001.

[SMBG01] L. Suhl, T. Mellouli, C. Biederbick, and J. Goecke. Managing and preventing delays in railway traffic by simulation and optimization. In M. Pursula and Niittymäki, editors, *Mathematical methods on Optimization in Transportation Systems*, pages 3–16. Kluwer, 2001.

[SS99] G. Schöbel and A. Schöbel. WabPlan – a software tool for design and evaluation of tariff systems, 1999.

[SS01] A. Schöbel and T. Sonneborn. Anschlusssicherung in multimodalen Verkehrssystemen. *Jahresbericht der Stiftung Innovation Rheinland-Pfalz*, 2001.

[SS03] A. Schöbel and M. Schröder. Covering population areas by railway stops. In *Proceedings of OR 2002, Klagenfurt*. Springer, 2003.

[SS05] A. Schöbel and S. Scholl. Line planning with minimal transfers. In *proceedings of ATMOS 2005*, 2005.

[Ste88] W. Stemme. *Anschlußoptimierung in Netzen des öffentlichen Personennahverkehrs*. PhD thesis, Technische Universität Berlin, 1988.

[SU89] P. Serafini and W. Ukovich. A mathematical model for periodic scheduling problems. *SIAM Journal on Discrete Mathematic*, 2:550–581, 1989.

[SV74] K. Sarkadi and I. Vincze. *Mathematical Methods of Statistical Quality Control*. Academic Press, New York-London, 1974.

[ta98] Metropolitan transit association. Urban fact book, 1998.

[Tam02] A. Tamir, July 2002. personal communication.

[TJ05] J. Törnquist and J.A.Persson. Train traffic deviation handling using tabu search and simulated annealing. In *Proceedings of HICSS'38*, Hawaii, 2005.

[TR73] C. Toregas and C. ReVelle. Binary logic solutions to a class of location problems. *Geographical Analysis*, 5:145–155, 1973.

[TSRB71] C. Toregas, R. Swain, C. ReVelle, and L. Bergman. The location of emergency facilities. *Operations Research*, 19:1363–1373, 1971.

[Tör05a] J. Törnquist. Computer-based decision support for railway traffic scheduling and dispatching: A review of models and algorithms. In *proceedings of ATMOS 2005*, 2005.

[Tör05b] J. Törnquist. N-tracked railway traffic re-scheduling during disturbances: theoretical and practical implications. Technical report, Blekinge Institute of Technology, Sweden, 2005.

[VD01] S. Voß and J. Daduna, editors. *Computer-Aided Transit Scheduling*, volume 505 of *Lecture Notes in Economics and Mathematical systems*. Springer, 2001.

[vE01] R.J. van Egmond. An algebraic approach for scheduling train movements. In *Proceedings of the 8th international conference on Computer-Aided Transit Scheduling, Berlin, 2000*, 2001.

[Voß92] S. Voß. Network design formulations in schedule synchronization. In *Computer-Aided Transit Scheduling*, volume 386 of *Lecture Notes in Economics and Mathematical Systems*, pages 137–152. Springer, 1992.

[Wag02] D. Wagner, editor. *Algorithmic Methods and Models for Optimization of Railways (ATMOS) 2002*, volume 66(6) of *Electronic Notes in Theoretical Computer Science*, 2002.

[Wag03] D. Wagner. Algorithms and models for railway optimization. In *Lecture Notes in Computer Science*, volume 2748, pages 198–206. Springer, 2003.

[War62] S. Warshall. A theorem on boolean matrices. *Journal of the ACM*, 9(1), 1962.

[WC74] J. White and K. Case. On covering problems and the central facility location problem. *Geographical Analysis*, 6:281–293, 1974.

[Wei81] W. Weigand. *Graphentheoretisches Verfahren zur Fahrplangestaltung in Transportnetzen unter Berücksichtigung von Pufferzeiten mittels interaktiver Rechentechnik*. PhD thesis, Technische Universität Braunschweig, 1981.

[Wei99] A. Weißler. *General Bisectors and their Application in Planar Location Theory.* PhD thesis, Universität Kaiserslautern, 1999.

[Wil99] N.H.M. Wilson, editor. *Computer-Aided Transit Scheduling*, volume 471 of *Lecture Notes in Economics and Mathematical systems.* Springer, 1999.

[Wol98] L. A. Wolsey. *Integer Programming.* Wiley, 1998.

[Wre81] A. Wren, editor. *Computer Scheduling of Public Transport.* North-Holland, 1981.

[WS05] S. Wegele and E. Schnieder. Dispatching of train operations using genetic algorithms. In *1st International Seminar on Railway Operations Modelling and Analysis*, Delft, 2005.

[XC94] X.Cai and C.J.Goh. A fast heuristic for the train scheduling problem. *Computers and Operations Research*, 21(5):499–510, 1994.

[Zar01] C. Zaroliagis, editor. *Algorithmic Methods and Models for Optimization of Railways (ATMOS) 2001*, volume 50(1) of *Electronic Notes in Theoretical Computer Science*, 2001.

[Zwa96] P.J. Zwaneveld. Routing trains through railway stations: Model formulation and algorithms. *Transportation Science*, 30:181–194, 1996.

Index